OPTIMUM FEEDING OF DAIRY ANIMALS

For Meat and Milk

Optimum Feeding of Dairy Animals

For Meat and Milk

MARSHALL E. McCULLOUGH

Charts and Illustrations by
George P. Joiner

The University of Georgia Press
Athens

To H. B. Monier who inspired his students to do more than their best, R. F. Elliott who instilled the idea of excellence, and the dairy farmers of Georgia whose friendship has been a constant inspiration.

First Edition 1969. Second Edition 1973.

International Standard Book Number: 0-8203-0239-2

Library of Congress catalog card number: 74-90556

Contents

List of Tables

List of Figures

Preface to First Edition

When one considers that there are some 20,000 new books published each year, the decision to add another to the list cannot be taken lightly. To justify a new volume, the author must recognize a need not currently being fulfilled. In my case, this need was for a single book which would bring together the underlying principles of nutrition, economics, and new techniques for calculating rations for dairy animals. The several books available on each subject provide excellent background materials, but the problem of putting them all together was still left to the initiative of the individual worker in the field.

A second factor was the need to present a new view of feeding dairy animals that included all available factors currently known to be important in this field. For nearly a century methods for calculating rations for ruminants have been based on tables of nutrient composition (many of which include values far removed from current feedstuffs) and tables of nutrient requirements which are all-too-frequently considered as absolute values. These systems have largely ignored the very important factors of ration digestibility and level of intake. My many years of experience in research and in trying to answer dairymen's questions in this field long ago convinced me that a new approach was needed. The ideas recorded here are not only supported by research but have been used almost daily in planning feeding programs on dairy farms.

The help and cooperation of many people who contributed to the preparation of this volume must be acknowledged. First, the author is grateful for the several administrators in the College of Agriculture, University of Georgia, who offered encouragement and cleared the way for its preparation. Secondly, the counsel and advice of my close associates was genuinely appreciated. Thirdly, a special note of thanks is due H. K. Welch and the Extension Dairy Science personnel for their time and effort in reviewing the original outline and the final manuscript. Finally, there was that efficient and hardworking secretary that always plays such an important role in the preparation of a book. In this case this role was ably filled by Mrs. Rosemary Ballard and it is to her that the author owes an eternal debt of gratitude.

Marshall E. McCullough

Animal Science Department
Georgia Station
University of Georgia

Preface to Second Edition

The world-wide acceptance of the first edition of this book has been most gratifying to the author. A Spanish translation, *Alimentacion practica de la vaca lechera,* was prepared by Julio Fernadez Carmona, and published by Biblioteca Agricola Aedos in Barcelona. The chapter on ration programming was included in Jorge De Alba's text entitled *Alimentacion del ganado en America Latina.* In addition to the adoption and use of the book by research and teaching personnel, it has enjoyed wide use by the feed industry and dairy farmers throughout the world.

The ideas and principles included in the first edition have been subjected to research efforts and extensive on-the-farm experience and have been found useful and sound. The availability of labor-saving feeding techniques and ration parameters for high levels of production are both timely and useful with the ever increasing demand for milk and dairy products.

The second edition has been prepared to include the major changes in the technology related to Optimum Rations which have occurred in the last three years.

The use of net and metabolizable energy is now universal. Thus, the tables listing feed ingredients have been completely revised to include this change. A complete least-cost linear program has been added as an appendix. This program has received extensive use in Georgia and surrounding states.

The primary addition is a chapter on meat production from dairy steers. The rapidly increasing demand for beef has resulted in this phase of nutrition becoming an important part of the dairy industry. To complete this phase, the tables of feedstuff composition and the least-cost linear program includes values applicable to meat production.

Finally, it is a pleasure to express my appreciation to all the people who made the first edition a success.

Marshall E. McCullough
Georgia Station
University of Georgia

Chapter 1

Some Philosophy and A Bit of History

"Agriculture is doubtless one of the oldest, most honorable and important pursuits among civilized nations. Without it the food of man must have been limited to the flesh of wild animals and the spontaneous productions of the earth: Commerce could not exist to any extent; the arts and sciences would be almost unknown; and society could not advance in improvement beyond a refined state of barbarism. But the culture of the soil enables men to produce more of the necessary food than they require, so that a part only are required in this pursuit, while the remainder are enabled to turn their talents and ingenuity to some other useful calling, the products or services of which are given to the agriculturists in exchange for food." This quote from the introduction to one of America's early books on agriculture must rank as one of the more eloquent statements of the role of agriculture in society. The book was entitled "Scientific Agriculture," its author, M. M. Rogers, a New York physician, and its publication date was 1848. This quote was chosen to introduce this chapter not to glorify the art and science of feeding dairy cattle but to set the stage for understanding the difference between just feeding cows and feeding cows as a part of the whole of social progress. Unless one understands the true place of agriculture in world progress and his role in agriculture, his job becomes merely something that is done to earn money for the necessities of life and his mediocre performance shows it. It is the understanding and real concern for the importance of one's work that separates the innovators from the followers of the world.

To most people, knowing something of the history of their profession gives a feeling of their being a link with the past and a step to the future. With this in mind, let us take a brief look at the people and events that brought the art and science of feeding dairy cows from the use of spontaneous productions of the earth to today's highly organized and highly productive state.

CATTLE THROUGH THE AGES

The beginnings of milk production, in common with all agriculture, is lost in the area of unwritten history. The role of the cow in the "scheme of things" is quite clear. The Creator set up the ruminant as an important link in the energy chain—energy from the sun is utilized by plants which are consumed by ruminants which are in turn consumed by non-ruminants, including man.

E. W. Shanahan in his book "Animal Foodstuffs" divided the era of recorded history of animal agriculture into three periods. Now we must add a fourth period to cover the growth of the dairy industry since World War II.

The first period is that marked in Greece, Italy, and neighboring countries, by the dominance of the Greek and Roman peoples for several centuries, both before and after the commencement of the Christian era. Agriculture and animal husbandry were held in high esteem by both these peoples. The existence in both Greece and Italy of mountain slopes close to populous plains favored the extensive rearing of sheep and goats for clothing material as well as for meat and milk, while the fairly advanced state of cultivation in the plains made it possible to raise a surplus of cattle for food above the requirements for draught purposes; but as the population increased, there must have been a relatively smaller quantity of meat available from such sources. At no time, however, do these peoples seem to have been great consumers of either meat or dairy products except maybe cheese. The general cultivation of the olive supplied an oil that was capable of taking the place of butter and animal fats used so freely at the present day in colder climates.

Throughout Europe, north of the Alps, the conditions during classical times naturally showed some differences as compared with the Mediterranean countries. Some domestic animals were reared on pastures, but not mainly for food purposes.

The period known as the Middle Ages may conveniently be taken as marking off a second stage. Before, as well as during this time, there was a continuous gradual shifting of the geographical center of European population northward from the Mediterranean, and westward toward the Atlantic. This period saw the rise of large and wealthy trading towns in Italy, Germany, France, the Netherlands, and Southern England, while the rural districts remained comparatively poverty-stricken. Hunting and warfare were the chief occupations of the nobility, who took little interest in agricultural pursuits.

In this society, the ox retained its traditional role as a beast of burden. As such, the ox was one of man's most valued possessions. Writing in the first century A.D., Columella reflected upon this importance and felt that the ox should be honored because ". . . it is still man's most hardworking associate in agriculture," and so great was the respect in which it was held among the ancients that it was equally a capital crime to have killed an ox and to have killed a fellow citizen.

Since the nobility owned the horses and they were primarily concerned with hunting and war, the horse was not yet a beast of burden and much of the labors of the ox went into providing feed for his eventual successor. This eventuality was foreshadowed by the invention of the horse collar and the traction harness somewhere around 1000 A.D. Simultaneously and perhaps as a result of these developments, there was a very gradual but constant increase in dairying. John H. Moore, in his article "The Ox in the Middle Ages," noted that the supremacy of the ox in agriculture was not seriously threatened until the nineteenth century. Oxen were listed in the United States census until 1890, and in that year there were 1,117,494.

Arguments of the relative merits of the ox and the horse as working animals undoubtedly began very early. The arguments were of little value until the invention of an iron plow that could withstand the spasmodic gait of the horse and the resulting thrusts against the hidden rocks. Thus another invention, the steel plow, foreshadowed the eventual demise of the ox as a beast of burden in Europe.

In addition to his role as draught power, the ox played another role that became increasingly more important. With the beginnings of the culture of crops, the ox became the main source of fertilizer for intensive agriculture. In the early stages of major agriculture, the medieval farmer was ignorant of the cultivation of hay. Thus the farmer was caught in a three-pronged dilemma. To plant his field he needed draught power. To make his grain grow he needed manure. Yet to get his manure he had to have animals which lived part of each year on the meager crops he raised. It was a vicious, unending struggle for oxen, manure, food, and land.

Beginning with the seventeenth century, a third stage in the development of European supplies of food can be distinguished. This period covered the era to about 1875 when cheap transportation and refrigeration entered the picture and changed forever the agricultural scene in the western world. The eighteenth century saw the introduction of rotation farming on a permanent basis, furnishing fodder and root crops as regular winter feedstuffs for livestock. It was during this period that horses came to be used as workstock and cattle were set free for the production of meat and milk for food. As the nineteenth century advanced, grain came to be transported more cheaply by land and sea, so that not only was land, once used for growing cereals, once more set free for animals, but animals also could be raised with the aid of imported materials on land that would otherwise have been incapable of supporting them. A frequently overlooked contribution to livestock improvement and the development of agriculture in this era was the gradual turning of the attention of landowners and the gentry from war to agriculture.

It was in this period that the United States was settled by peoples from England and Western Europe. In addition to their cattle, they brought with them European methods of farming and European crops.

The fourth era in terms of the dairy industry in the United States began about 1850 and extends to the present. In 1850 the census listed 6,385,094 cows kept for milk, many of which were congregated around large cities. This number increased gradually until 1890 when 12,443,120 cows were listed. Here we enter the great period of growth in the dairy industry of the United States made possible by transportation and most of all by refrigeration. Too much credit cannot be given to refrigeration in the development of the dairy industry since it freed the dairy farmer from a close proximity to large cities and made possible the development of small dairies through the land. The next thirty years, (1880 to 1910) saw cow numbers increase to 20,625,432. The increase in numbers was small from this date onward and reached a peak in 1940 with 24,397,000. Since World War II, cow numbers have constantly declined.

Just as the period from World War II to the present has been a period in which cow numbers have declined sharply, it has also been the era of greatest advancement in the application of technology. In figure 1, the average milk production of cows enrolled in DHIA (Dairy Herd Improvement Association) is traced from 1906 to 1966. The data show that it required 54 years for the average milk production of 5,300 pounds in 1906 to double. The nearly 2,000 pound increase between 1960 and 1966 is equal to the total improvement between 1906 and 1930.

CROP INTRODUCTION AND AMERICAN DAIRYING

In every country in every age, farmers have shown great interest in growing new crops or new varieties of old crops imported from another country or another area of the same country. This interest probably reflects their continuing search for a means of improving their lot through new

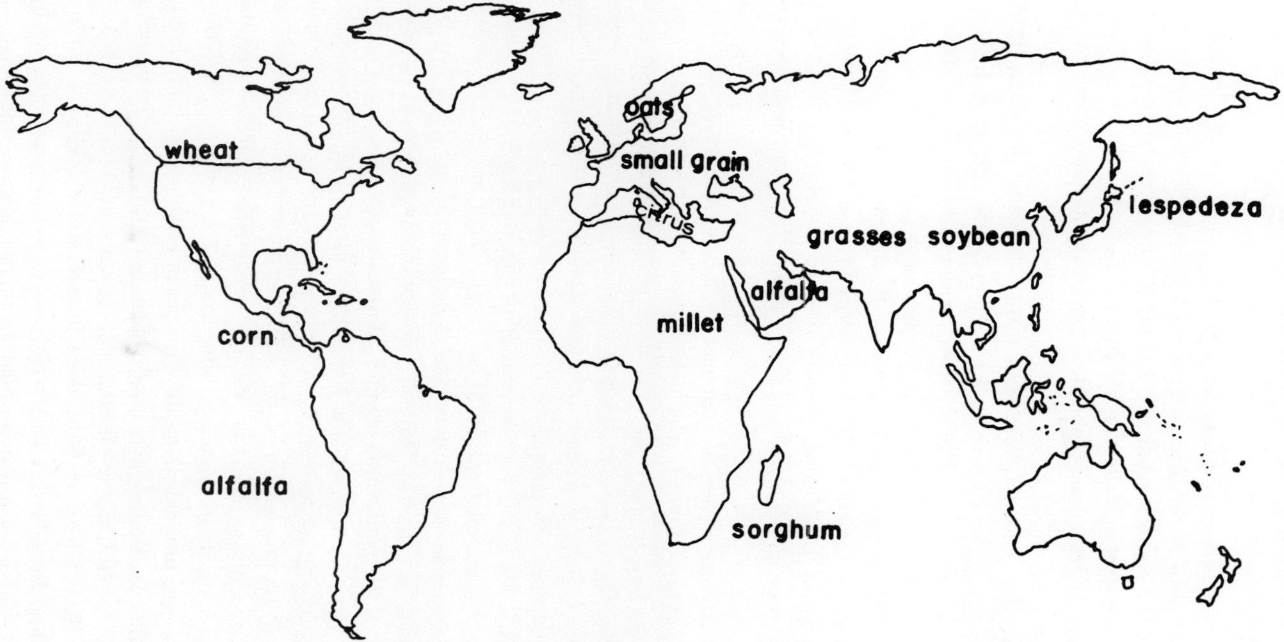

Figure 1 The Origin of Several of the Important Crops which have Contributed to the Development of the American Dairy Industry.

crops or better yields of old crops. In no area of the world has this tendency contributed more than in the borrowing of the crops of the world by American agriculture. The origin of many of our importations are shown in figure 2. Due to the great expanse of the United States, climates and soils suitable for a broad range of crops made the list of usable importations greater than in most countries.

Of the crops that were to play a significant role in the development of the dairy industry in the United States, corn was the only crop present when the early settlers came. Due to their European heritage, the settlers either brought with them or imported very early such grains as wheat, oats, barley, and rye. These crops were continuously imported and exported as new introductions or varieties, and the exact date or origin of current varieties can hardly be established. Among the prominent early Americans who imported were Benjamin Franklin, Thomas Jefferson, and George Washington.

Of special interest to the development of the dairy farms of America was the importation of forages. The native grasses in the areas that were to become important as dairy farming areas were of little value as feed crops. Following the American Revolution various kinds of forage seeds such as ryegrass, red clover, white clover, and sainfoin began to appear more frequently in European shipments to the United States.

Timothy as a hay grass became one of the most important cultivated grasses during this period from 1820 to 1860. There are many discrepancies in the accounts of its history, but it is generally accepted that the grass was named for Timothy Hanson, who cultivated it first in Carolina in the middle of the eighteenth century. A peculiar interest has been attached to this grass since it was the first to be cultivated in America. The contribution of this grass to feeding of dairy cows in America is difficult to overestimate. Its dominance of the hay crop lasted until the development and widespread use of alfalfa. Many writers credit Columbus with the first introduction of alfalfa although little came of his efforts. Alfalfa, of Arabic origin, was introduced to the American colonies by Eliza Lucas again with little results. Its rapid development in the United States dates from the 1850's when it was re-introduced from Chile and known as Chilean clover. Even the Chilean variety was not sufficiently hardy to survive northern winters. Introduction from colder regions and improvements by selection and breeding eventually resulted in alfalfas that are able to survive the winters of any area in the United States.

Italian ryegrass, a perennial, was introduced into the United States in the 1830's from Europe.

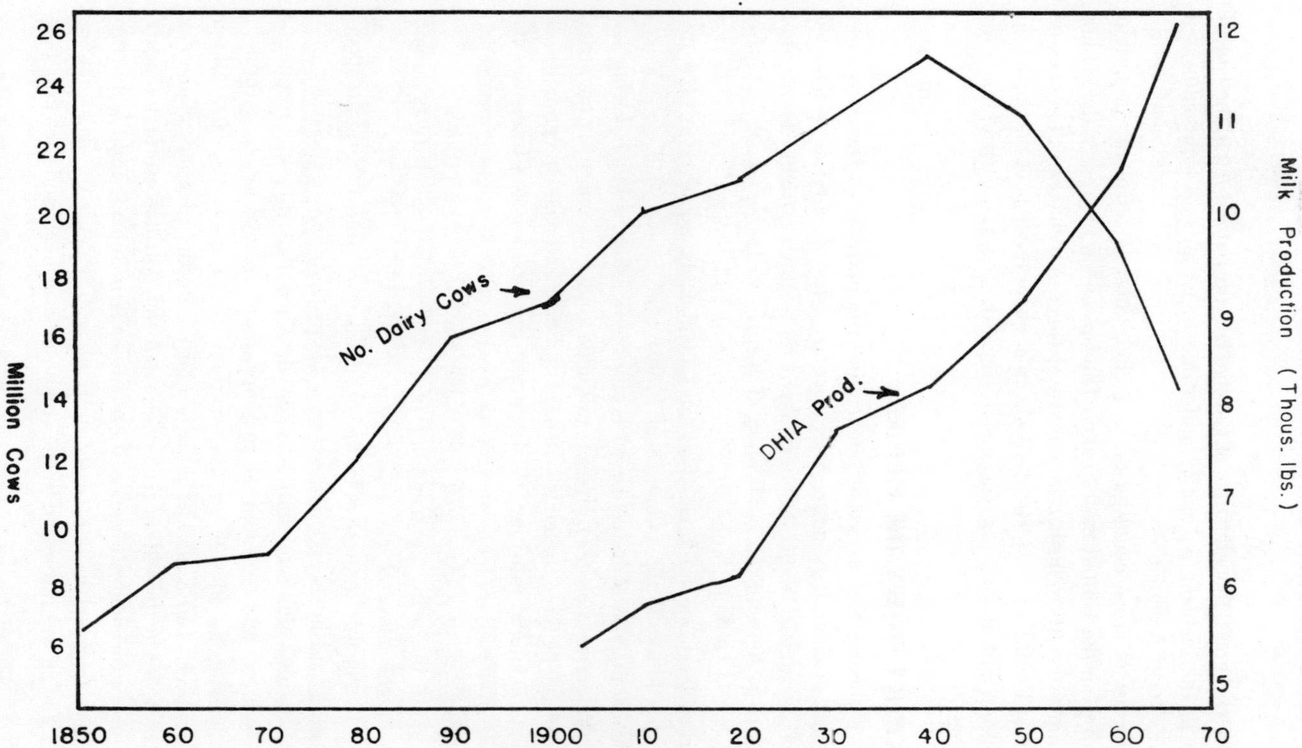

Figure 2 The Changes in Dairy Cow Numbers and the Increase in Milk Production per cow in DHIA herds in the United States.

Orchard grass, though native of Europe, became naturalized in the United States in the eighteenth century and came to be an important cultivated crop in North America.

The list of crops could go on to include most or all of the important feed crops in the United States today. The succession of crops reflected not only improvements in husbandry but also changes in technology. The records show that Franklin introduced the soybean at about the time of the American Revolution but it was not successful since there was no machinery for handling it.

SCIENCE ENTERS THE PICTURE

In reviewing the history of science and the nutrition of the dairy cow, the initial problem is deciding where to start. Most discoveries in the field of biology exerted some influence on dairy cattle nutrition. The author prefers to begin with the publication of Justus von Liebig's book "Organic Chemistry in Its Applications to Agriculture and Physiology" in 1842. This date is at least one hundred years later than many would begin, but this is the beginning of the influence of Liebig who, as we shall see, exerted an unusual amount of influence upon all of agricultural science. Liebig and his laboratory at Giessen, Germany, attracted students from many countries. Included among these students were most of the men who laid the groundwork for modern animal nutrition. A significant contribution of Liebeg was the practice of having students work in the laboratory on actual problems rather than simply sit in a classroom and watch teacher demonstrations.

Among the many contributions of Liebig to the chemistry of animal nutrition, the following are outstanding.

1. Liebig was interested in the phenomena of animal respiration and body work and in the relation between combustion and respiration.

2. Liebig's differentiation between the food that supports respiration (carbohydrates and fats) and the food apparently needed for work (protein, or, expressing the idea in another manner, the necessity for extra protein for hard work) had a marked influence on the dietary practices of the day. This work led to two ideas that dominated the nutrition world for nearly a century: a very high allowance of protein for farm animals and the nutritive ratio that appeared in most early nutrient requirements for dairy rations.

3. Liebig was interested in the composition of foods and their functions in the body.

4. Although aimed primarily at plant nutrition, Liebig's emphasis on the "law of the minimum" contributed greatly to research development and extension teaching in the late nineteenth and early twentieth century. The idea was not original with Liebig but his use of it resulted in its being called Liebig's law. The idea of biological processes being controlled by the element in least supply led to exhaustive testing of soils, plants, and feedstuffs and the establishment of tables of chemical composition for formulating rations.

5. The research and writings of Liebig resulted in the establishment of agricultural experiment stations. The first station to exert a major influence on animal nutrition was established in Moeckern, Germany in 1852. Among the statutory provisions governing the new station was a provision for investigating the following: "The ingredients of plants and their actions on the animal organism. The various kinds of feeds, their composition and their valuation for the production of flesh, milk, wool, etc." and "The construction of authentic tables of comparative numbers, having reference to all branches of agriculture, as for example, tables of the relative value of feed."

Another significant German contribution was the introduction in 1860 by Henneberg and Stohmann of the Weende method for fractionating feed into water, ether extract, crude protein, crude fiber, ash and the computation of nitrogen-free-extract by difference. This method has been under almost constant criticism since its introduction, but its use is still world-wide a hundred years later.

The full weight of the German work hit American agriculture and animal nutrition in a now classic series of papers published between 1875 and 1880. One of these papers that deserves special attention was prepared by Professor W. O. Atwater as part of the Report of Work of the Connecticut Agricultural Experiment Station for 1877–78. Since this work had so great an effect on American agriculture and the original manuscript is not readily available, it seems worthy of extensive quotation. The author had received a copy of Wolff's book covering some fifteen years of German research and his manuscript contained many quotations from the German work.

"I have before me a book by Wolff entitled 'Nutrition of Animals Useful in Agriculture', in which are gathered together the results of the experiments upon this subject during the 15 years previous to 1876. It is an octavo volume of 552 pages. It gives the general plans and outlines of the results of probably 1,500 feeding experiments, each one of which was carried out with a thoroughness never equalled by any feeding experiment on this side of the Atlantic ocean."

"There are two very important matters connected with the economical feeding of stock which the teaching of modern science explains, but which too few farmers understand. First, how to adapt the food most economically to the wants of the animal and the purpose for which it is fed; and second, how to feed so as to make the richest manure."

The listing of these two items indicates two of the more important problems of the livestock industry in the last half of the nineteenth century. The first item still applies today; but, the continuing problem of having manure as a fertilizer disappeared with commercial fertilizers and this is one of the last references to the production of manure as a part of animal nutrition.

Professor Atwater proceeded to outline what he considered to be the fundamental principle in economical feeding.

"The right feeding of stock, then, is not merely a matter of so much hay and grain and roots, but rather of so much water, starch, sugar, gluten, etc., of which they are composed. To use fodder economically, we must so mix and deal it out that the ration shall contain just the amounts of the various ingredients needed for maintenance, and for the particular form of production that is required.

"To this matter squarely before us, we must consider: (1) What is the chemical composition of our fodder materials? How many pounds of albuminoids, sugar, starch, fat, water and other ingredients are contained in a hundred pounds of hay, clover, potatoes, meal, etc.? (2) Of these various ingredients of food, what proportions of each are digestible and consequently nutritious? (3) What part does each of these food ingredients play in the animal economy? Which are the "flesh-formers", i.e., make the muscle; which ones are transformed into the fat (butter) and casein (curd) of the milk; from what ones are the heat and forage produced? (4) How much of each do different animals, as oxen and cows, need for maintenance of life and production of meat, milk, etc.? And finally, how must different kinds of fodder be mixed and fed so that the digestible material shall be most fully digested and utilized, and the least quantity wasted?"

Most students of nutrition would agree that Professor Atwater stated the problem facing research in nutrition quite well. The two major items that were missing were vitamins and minerals. In addition to the discovery of vitamins and the establishment of requirements for minerals, the major results of research in nutrition since 1878 have been in supplying some of the answers to the four questions posed above.

In addition to the above discussion of fundamentals, Professor Atwater commented on several feeds that had been studied, reproduced Wolff's first feeding standard and attached a long table summarizing the results of chemical analysis and digestion trials with feedstuffs both in Germany and at the Connecticut station.

The brief summary of the growth of the dairy industry and the science of nutrition has been cut off at the end of the nineteenth century. The remainder of this book will be devoted to the findings of the twentieth century. So brief a summary of so much history must omit many important events and the names of many people who made important contributions. The author will have satisfied his purpose if the reader now possesses two things: (1) an appreciation for the many events that must take place to enable an industry to progress and, (2) some appreciation of the heritage that is ours as we become the link of the dairy industry between the generations past and generations yet to come.

References

1. Atwater, W. O. Report of Work of the Agricultural Experiment Station, Middletown, Connecticut. 1877–78.
2. Kerr, Homer L. Introduction of Forage Plants into Ante-Bellum United States. Agricultural History 38:87. 1964.
3. Moore, John H. The Ox in the Middle Ages. Agricultural History 35:90. 1961.
4. Moulton, F. R. Liebig and After Liebig. A Century of Progress in Agricultural Chemistry. AAAS, Washington. 1942.
5. Rogers, M. M. Scientific Agriculture. Erastus Darrow, Rochester. 1848.
6. Shanahan, E. W. Animal Foodstuffs. George Routledge and Sons, London. 1920.

Chapter 2

The Language of Nutrition and Guides to Feeding Practice

During the last hundred years many attempts have been made to devise systems which would permit farmers to compute, for their cattle, rations which would supply the nutrients needed to permit growth, reproduction, maintenance and milk production. The systems consist of two parts: first, tables showing the composition, digestible nutrients and estimated feeding value of feedingstuffs; and second, statements of the nutrient requirements of animals of different size, age and condition, when growing, fattening or milking at different rates.

In 1809, Thaer, a German investigator, set himself the task of establishing a table of equivalents which would indicate the relative feeding values of different feeds in terms of a common unit. At that time good meadow hay was universally regarded as a complete feed, suitable for practically all purposes, and it was natural that Thaer chose this as his standard of comparison. Thus arose the concept of "hay values," the sum of all the soluble ingredients of the hay being taken to represent its nutritive value. The sum of the soluble ingredients of all other feedstuffs were used as a basis of comparison and their "hay value" was computed as a comparative value. Twenty-five years later, in 1836, Boussingault drew attention to the importance of protein as a constituent of the rations of livestock, and modified Thaer's tables by calculating the hay values on a protein basis.

In 1864 two other German workers, Henneberg and Stohmann, demonstrated that the former tables were misleading since they were based on the assumption that all the crude nutrients of the food could be utilized by the animal.

Henneberg and Stohmann, by measuring both the intake of feed and the amount of nutrients excreted in the feces, were able to show that part of the food was excreted and was therefore useless to the animal. This important advance formed the foundation for the modern methods of computing the nutritive value of feeds in terms of digestible constituents instead of crude constituents. Too much emphasis cannot be placed on the importance of the work of Henneberg and Stohmann since their experiments established not only the concept of digestibility but also laid the foundations for the "balance" experiments which have played such a significant part in modern nutritional research.

THE LANGUAGE OF NUTRITION

In all human activities, it is necessary for people to communicate. In nutrition this led to the development of a language that must be understood before one can utilize the tools and knowledge of the field. A major problem in the field has been the lack of uniformity in the meaning of words and in the units of measure used to express nutritional relationships. In an effort to unify the usage of terms and, particularly, to have everyone using the same units of weights and measures, two standards have been adopted by most workers around the world. The basic unit of measure for weight is the metric system. Detailed tables of the metric system may be consulted in any textbook of mathematics. The only units that must be understood here are those relating to weights of feed. The basic unit is the **gram** usually written simply as **g.**

1 kilogram (kg) = 1000 grams (g) = 2.205 pounds (lb)
1 lb. = 453.6 g.
1 oz. = 28.35 g.

To convert:	Multiply by:
lb to kg	0.4536
kg to lb	2.205

The second set of terms that must be understood is the language of energy.

The basic unit is the **calorie** usually written as **cal.** As usually used in nutrition, a small calorie is the amount of heat required to raise the temperature of one gram of water from 14.5 to 15.5 degrees centigrade. The **c** is always written as a small **c** to differentiate it from Calorie spelled with a capital **C** (1,000 small calories) under the older usage.

Kilocalorie (kcal) A kilocalorie is 1,000 small calories. Kilocalorie is preferred to Calorie because of the difficulty in deciding whether the **C** is small or large.

Megacalorie (Mcal) A megacalorie is equivalent to 1,000 kilocalories or 1,000,000 calories. A megacalorie is equivalent to a **therm (T).**

The data in figure 3, illustrate the general relationship between the terms used to describe the measures of energy.

Gross energy (GE) is the amount of heat, measured in calories, that is released when a substance (in this case, feed) is completely oxidized in a bomb calorimeter under prescribed conditions. This value has no nutritional significance but is necessary as a starting point in calculating the remaining measures.

Digestible energy (DE) is the feed intake gross energy minus the amount of energy in the feces. Since fecal energy contains a small amount of energy other than from the feed, the value is sometimes referred to as apparent digestible energy.

Metabolizable energy (ME) is the food intake gross energy minus fecal energy, minus energy in the gaseous products of digestion, minus urinary

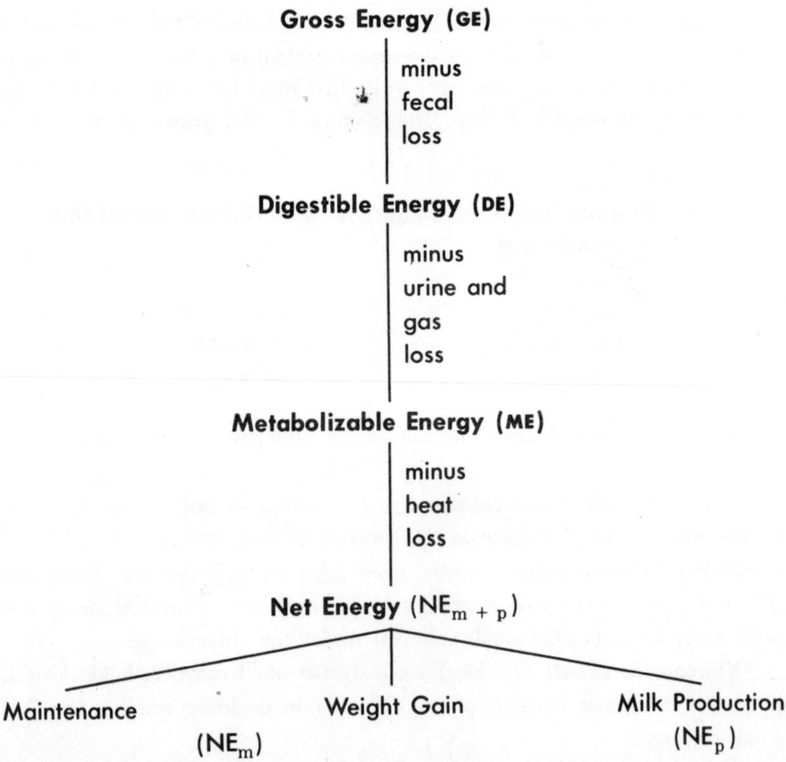

Figure 3 The Utilization of Energy (Conventional system).

energy. This value differs from digestible energy in that the additional losses of energy in the gases formed during digestion and the energy excreted through the urine are also subtracted from the gross energy.

Net energy (NE) is the gross energy consumed minus fecal energy, gaseous energy, energy in the urine and energy expended in heat production during fermentation and nutrient metabolism. As will be seen later, net energy is commonly divided into that used for maintenance and that used for production. For this reason, total net energy is written as NE_{m+p}—that used for maintenance as NE_m and that for production as NE_p.

Another term frequently used in research and sometimes in popular literature is **metabolic size.** The term is frequently written as ${W_{kg}}^{0.75}$. This means that the body weight of the animal is stated in kilograms raised to the 0.75 power. This measure is used in recognition of the fact that nutrient requirements of animals vary with the size or surface area of the animal which is not fully represented by body weight. In other words, a 1,600 lb. cow is not twice as large as an 800 lb. cow, although she weighs twice as much. In practice, metabolic size has little practical meaning in terms of cows of the same breed in the same herd and the term is best left to the use of research workers.

The widespread use of statistics in research has introduced many new terms that are frequently used without explanation. There are two items from statistics that should be understood by all people who use nutritional information.

The first term is **significance.** When two or more rations are fed to animals and a difference in animal performance (milk production, weight gain, etc.) is measured, the question arises as to whether or not this measured difference is due to the rations fed or to some other factor which affected the performance of the animals. Research workers are by nature conservative creatures. Thus most research data are subjected to mathematical treatments called statistics. In its most simplified form this mathematical treatment is a means of calculating the odds that the differences measured in experiments have of being due to the rations fed and not to chance. The two most frequently used measures are to state that an observed difference is significant at the 0.05 or 0.01 level of probability. This means simply that at the 0.05 level there are 95 chances out of 100 that the measured difference is due to the rations fed and not due to chance. Similarly the 0.01 level indicates that the odds are 99 to 1 or only 1 chance out of 100. A point that should be made is that requiring such odds before calling a measured difference significant does not mean that a difference that had odds of say 90 to 10 is not a true difference. The insistence by research workers on such high odds before calling measured differences **significant differences** is a recognition of the fact that there are so many factors which affect animal performance that small differences due to rations must be carefully evaluated before they are recommended for general usage.

A second statistical measure that should be understood is the **regression equation.** This method of presenting the results of research and of expressing nutritional relationships is so universally used that its meaning must be understood. The following regression equation resulted from research at the Georgia Experiment Station.

$$Y = 0.28X_1 + 9.5X_2 + 2.01X_3 \qquad R^2 = 0.85$$

Where: Y = Total TDN intake (lb)
X_1 = pounds of 4% milk produced
X_2 = body weight in pounds
X_3 = gain in body weight in pounds

The question being answered by the equation was "What determines the amount of TDN a cow will consume?" The equation indicates that in this case for each pound of 4 per cent milk the cows produced, they consumed 0.28 lbs. TDN; the cows consumed .95 lbs. TDN per each 100 pounds of body weight; and the cows consumed 2.01 lbs. TDN for each pound that they gained in body weight. Thus the daily intake of an individual cow would be the sum of the three items. The R^2 of 0.85 indicated that 85 per cent of the variation in TDN intake observed in the cows under study was accounted for by the three variables. The remaining 15 per cent of the observed variation was not accounted for.

A more common type of regression equation contains a constant indicating that the regression line does not pass through zero. Such an equation is illustrated by the commonly used one for converting TDN to ENE.

$$ENE = 1.393 \times TDN - 34.63$$

To use the equation the per cent TDN in a feed is multiplied by 1.393 and 34.63 is subtracted from the product. The constant is either added or subtracted depending upon the sign preceding it.

GUIDES TO FEEDING PRACTICE

With the general explanation of the language of nutrition "under our belt" we can proceed to a consideration of the accepted guides to feeding practice. These guides are frequently referred to as "Feeding Standard" or "Nutritional Allowances" but the author prefers the term "Guides to Feeding Practice" for reasons that will become clear.

As the experiment stations developed tables of composition of feeds and established other tables of comparative values, the next logical question asked by feeders was "how much should I feed?" In attempting to answer this question, it became obvious that some system of feed evaluation other than "hay value" was necessary. The three major systems which were developed were the Total Digestible Nutrient (TDN) system in the United States, the Starch Equivalent System (SE) in Middle Europe, and the Feed Unit (FU) in the Scandanavian countries. The Net Energy (NE) system developed by Armsby and others in the United States never became a major system largely due to a lack of a large number of feeds with net energy values determined.

It is not the author's intention to discuss here the research techniques used in determining the various measures of feed value. Rather we shall compare the systems on the basis of what they propose to measure. One thing all systems have in common is that all begin with digestion trials to measure the apparent digestibility of the crude protein, crude fiber, ether extract and nitrogen-free extract.

Total Digestible Nutrients (TDN) The present method of calculating TDN arose from combining caloric values for human foods with data for digestible nutrients in cattle. As we have seen, Atwater published an extensive report on European investigations with cattle which brought German feeding standards to the attention of American workers. Some years later, Atwater proposed the use of Rubner's factors for calculating the potential energy value of nutrients—namely, 4.1 kcal per gram for carbohydrates, 4.1 kcal for protein and 9.3 kcal for fat. The factors for carbohydrates and fat were gross caloric values, but the one for protein was corrected for the portion of the gross energy which was lost in the urine. The long road from this suggestion to the adoption of TDN by Henry and Morrison in 1915 has been reviewed by Maynard in his paper "Total Digestible Nutrients as a Measure of Feed Energy." At the end of the gradual process, the formula which has been in use arose:

$$\text{TDN} = \text{digestible protein} + \text{digestible carbohydrate} + \text{digestible fat} \ (\times\ 2.25).$$

From the very beginning two facts have been true for the TDN system: (1) it has been under constant attack as being of little value in evaluating feeds and, (2) most American workers have used the method both in research and teaching.

These two seemingly contradicting facts would appear worthy of some explanation. From the beginning, the weaknesses of the TDN system have been understood. At the same time, workers in animal nutrition have also understood that any measure of the nutritive value of a feed applies only to the exact conditions under which the value was determined. For this reason, no exact table of feed energy values could be prepared. Since TDN was the simplest and cheapest of all measures, its use multiplied.

There appears to be general agreement that TDN is a useable system for evaluating grains and in calculating grain rations but most agree that one of the net energy systems should be used in evaluating forages. This difference between feeds is, thus, largely due to the crude fiber fraction and its overall influence on feed utilizations.

The Net Energy Systems (NE) Instead of evaluating feeds simply on the basis of their digestible nutrients, several methods have been proposed for evaluating the actual ability of feedstuffs to contribute to the production processes in the animal. The three prominent systems are Kellner's Starch Equivalent, Armsby's Net Energy, and the Scandinavian feed unit based on barley as a standard.

Blaxter (*Energy Feeding Standards for Dairy Cattle*) illustrated the three systems in terms of their use for calculating the energy value of hay, corn, and peanut meal.

Armsby: Digestible organic matter lbs. ×
(hay, 1.588) (hay, 47.4)
(corn, 1.769) therms, less heat increment as (corn, 58.3)
(PM, 1.996) (PM, 52.6)
per 100 lb. dry matter = NE in therms.

The factor used to calculate the first part of the equation separated the three feeds by indicating the higher energy value of the corn and peanut meal due to their higher fat contents.

Kellner: Digestible true protein × 0.94 (hay, 1.91)
+ digestible ether extract × (corn, 2.12)
(PM, 2.41)

+ sum of digestible NFE and crude fiber = Crude S.E.

This value was corrected for crude fiber to give the starch equivalent as a percentage.

The Kellner system considers three items: (1) the protein fraction is handled separately, (2) the digestible ether extract was assigned a caloric value in keeping with the nature of the fat in each feed, and (3) the final value is corrected for the crude fiber content of the feeds.

The feed unit system is essentially the Kellner system with the factor 1.43 used for protein instead of Kellner's 0.94.

The data in table 1 illustrate the estimates of the nutritive value of two forages by the three systems. While the data indicate a need for using a net energy system to evaluate forages, they do not indicate a superiority for either of the systems.

In fact, the one thing that must be realized in using systems for evaluating feeds is that a feedstuff has no constant nutrient value. The important factors which influence the nutritive value of a feedstuff are:

1. The associative effects of other feeds in the ration. Forbes commented "an individual foodstuff expresses its normal and most characteristic nutritive value . . . only as it is a part of a ration which is qualitatively complete and quantitatively sufficient, for the conditions existing."

2. The plane of nutrition (the level of feeding as an increment of maintenance). Rations fed below or near maintenance exhibit a higher net energy value than rations fed at 2 × maintenance or above.

Table 1 Estimates of the nutritive value of an artificially dried grass and of a mature rye grass hay as assessed by application of factors to the digested proximate principles the feeds contain.[a]

Method of assessment	Units	Dried grass, 19% crude protein	Rye grass hay, 6% crude protein	Superiority of dried grass
				(%)
TDN system (America)	lb/100 lb.	64.1	49.6	29
ENE system (America)	Therms/100 lb.	53.4	36.6	46
Kellner's starch equivalent system (Germany)	kg/100 kg.	56.2	32.1	75
Modified starch equivalent system (United Kingdom)	lb/100 lb.	56.2	38.5	46
Fodder unit system (Scandinavia, France, Israel)	FE/100 lb.	80.4	43.2	86

[a] The factors employed are listed in reference 17.

3. The animal requirements for which the ration are fed. Rations have a higher net energy for maintenance than for gain in weight or milk production.

The author is in general agreement with the trend in the nutritional world to use one of the energy systems for evaluating feeds. The point to be stressed however, is that there really is no single value for a feedstuff and no system of evaluation is, as yet, a simple matter of arithmetic. In practice, rations should be calculated not only on the basis of available nutrients but also on the basis of a knowledge of the total ration.

The results of the thousands of digestion trials which have been performed during the last hundred years are available in tables in textbooks on feeding farm animals. Since the data are presented in terms of the method of evaluation used in that country, it is frequently desirable to calculate from one standard to another. The following equations may be found useful in such transformations.

1. Digestible energy to metabolizable energy.
   M.E. = 82% of digestible energy
2. Digestible energy (kcal/gm dry matter) from total digestible nutrients
   DE = −.546 + 0.055 TDN
3. Net energy (therms) from TDN
   Net Energy = 1.454 TDN − 38.82
4. Total digestible nutrients from starch equivalent.
   Per cent TDN = 0.777% S.E. + 0.127% crude fiber + 19.01 (concentrates)
   Per cent TDN = 1.079% S.E. + 0.483% crude fiber −1.09 (green forages and silage)
   Per cent TDN = 0.998% S.E. + 0.514% crude fiber +2.64 (dry roughages)
5. Starch equivalent from TDN
   Per cent S.E. = 1.035% TDN − 0.352% crude fiber −4.50 (concentrates)
   Per cent S.E. = 0.870% TDN − 0.398% crude fiber +1.47 (green forages and silages)
   Per cent S.E. = 0.986% TDN − 0.588% crude fiber +4.40 (dry forages)

NEW FEEDING SYSTEMS

The need for new ideas and systems for feeding livestock is clearly evident from the above discussion. This is particularly true as we intensify our livestock production. When we were dealing with small dairy herds of say twenty cows, each cow could be handled as an individual. If one of the feeding systems was used to outline a feeding program for the herd, any deficiency in the program could be corrected on an individual cow basis simply by the feeder observing the response and needs of each cow and adding a little more feed to some cows and taking away some from others. In large herds handled with a minimum of individual attention, this "eye of the Master" type of feeding is no longer possible. In recent years research workers have given much attention to this problem and three new proposals have been made that deserve attention.

The first system is the proposal put forward by Blaxter and adopted by the British Research Council in "The Nutrient Requirements of Farm Livestock—Ruminants." The proposed system takes into account the following factors: (1) the efficiency with which a diet is utilized increases with the concentration of energy per unit weight of the ration, (2) the metabolizable energy value of a ration is corrected for differences in plane of nutrition, (3) the utilization of the ration is influenced by the type of animal performance required of the animal.

The data in figure 4 illustrate Blaxter's preferred values for the efficiencies of utilization of metabolizable energy for maintenance, body gain and milk production. Beginning with a low concentration of metabolizable energy (1.6 Mcal/kg) typical of a ration consisting of low quality forage and proceeding to a very high concentration of 3.4 Mcal/kg (typical of an all concentrate ration), the general effects of ration energy concentration on animal performance can be seen. The concentration of energy in the ration has little real effect on the use of the ration for maintenance. This is in keeping with farm experience that lower quality feeds may be used to maintain animals so long as the level of intake is sufficient and the ration does not lack essential nutrients. The overall relationship between the concentration of energy and milk production is curvilinear. Increasing the concentration from 1.6 Mcal/kg to 2.2 Mcal/kg results in improved utilization. Between concentrations of 2.2 and 3.0 Mcal/kg the relationship is linear. Further increases in concentration results in decreased efficiency of utilization of energy. Again this general relationship is in keeping with farm

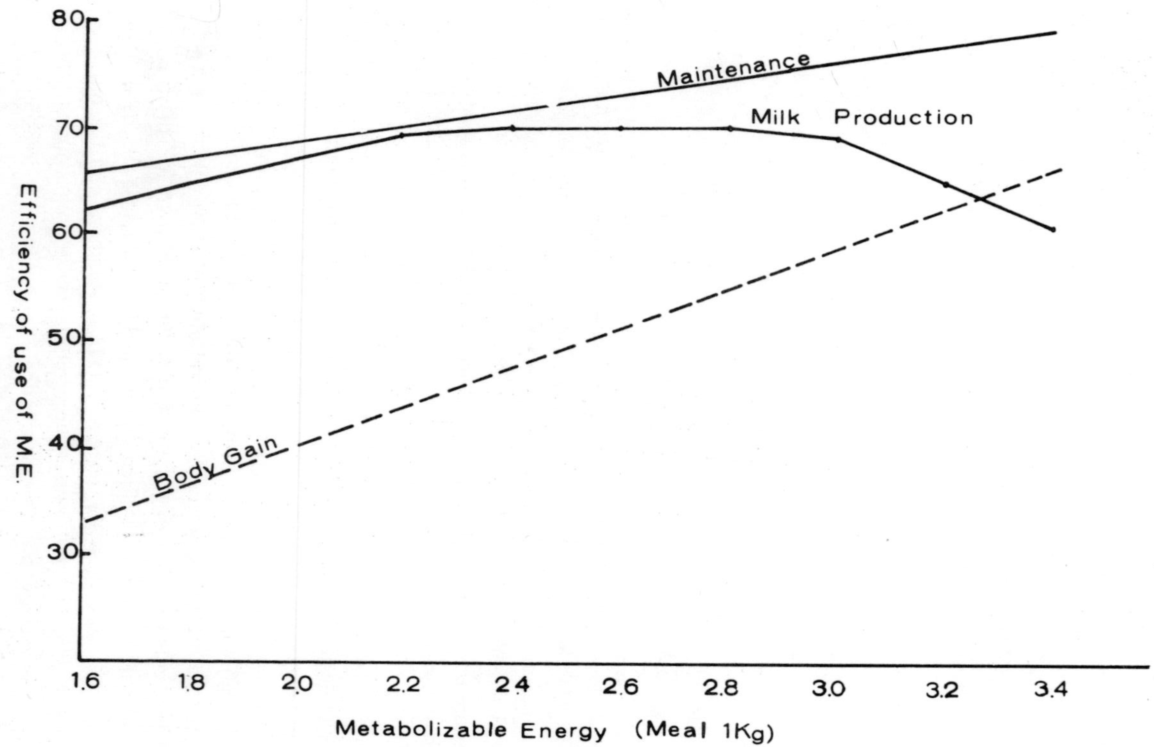

Figure 4 The Influence of Energy Concentration in the Ration and the Purpose for which the Ration is Used on the Efficiency of Energy Utilization.

observations and the results of feeding trials around the world. The relationship between energy concentration and the efficiency of energy utilization for fattening is the most dramatic. Again the relationship is supported by experience in feed lots and the very high efficiencies of fattening of steers which have been achieved on all concentrate diets. This system of stating the nutrient allowances for livestock is the first to bring scientific feeding in line with the "art" of feeding that has been practiced by leading feeders for many years. The system recognizes, as they have, that there are maintenance rations, fattening rations and milking rations.

The proposed feeding allowances for a mature dairy cow weighing 500 kg (1100 lbs) giving 4 per cent milk and walking one mile per day are in table 2. The increased efficiency of the rations with increasing concentrations of energy can be seen in the decreasing quantities of feed required

Table 2 Daily requirements of a mature cow weighing 500 kg., giving milk of 4 per cent fat content and walking one mile.

Concentration of diet (kcal metabolizable energy/g dry matter)	Milk yield						
	kg/day lb/day	5	10	15	20	25	30
		(Mcal metabolizable energy/day)					
1.8		19.1	28.6	—	—	—	—
2.2		17.7	24.9	33.3	—	—	—
2.6		16.8	23.0	29.8	37.2	45.4	—
3.0		16.3	22.4	28.7	35.2	42.1	49.3
3.4		16.5	22.9	29.5	36.1	43.0	49.9
Estimates of requirements by							
Evans [a]		18.0	23.6	29.2	34.7	40.2	45.8
N. R. C. [b]		19.5	25.2	30.9	36.7	42.3	48.1
Axellson [c]		20.0	25.5	30.9	36.4	41.8	47.3

[a] Evans (1960). The primary standards are 2.90 kg starch equivalent for maintenance and 0.26 kg starch equivalent for the production of 1 kg milk. The net energy value for fattening of 1 kg starch equivalent was taken to be 2.36 Mcal and the average efficiency of utilization of metabolizable energy for fattening 55%. 1 kg starch equivalent is thus equivalent to 4.28 Mcal metabolizable energy.

[b] N. R. C. (1958). The primary standards are 3.86 kg of total digestible nutrient for maintenance and 0.32 kg/kg milk. The metabolizable energy value of 1 kg total digestible nutrient was taken to be 3.58 Mcal.

[c] Axellson (1949). This standard is expressed in terms of metabolizable energy.

to produce each level of milk production as the energy in the diet increases from 1.8 Mcal/kg to 3.4 Mcal/kg feed.

This system has another very useful possibility particularly in terms of machine calculating total rations since the dietary concentrations of protein and minerals can be stated in terms of the concentration of nutrients in the dry matter. Table 3 illustrates this possibility in terms of a 500 kg cow during her dry period and at two levels of production.

As this system is subjected to further research and refinement, it has great possibilities and may replace the older starch equivalent and TDN systems.

Two American-developed systems for net energy have received considerable use and have replaced TDN in NRC Recommended Allowances. The first was developed by Lofgreen and Garrett at the California Station for growing and fattening steers and heifers. This system recognizes the difference in use of net energy for maintenance and production and divides this portion into NE_m (maintenance) and NE_g (gain). The system has been adapted by several nutritionists working with feed-lots and represents a distinct improvement over former systems. In practice some difficulties have arisen from the use of two values when computers have been used to calculate a single ration for a range in body weights with similar rates of desired gain. This problem has been satisfactorily handled by establishing ration characteristics in terms of NE_g only.

A second system has been developed by Moe and Flatt using a combined production value called $NE_{lactation}$ or NE_{milk}. Their system adjusts the various requirements of net energy for maintenance, growth, reproduction, and milk production into a single value. This system has resolved some of the problems associated with former recommendations and has worked well when properly applied in the field.

Both net energy systems suffer from the old problem in all feeding standards that the utilization of energy is not constant. Some of the factors affecting energy utilization are:

1. Nutrient balance of the ration in relation to the requirements of the animal
2. Level and ratio of nutrients fed

Table 3 The nutrient requirements of an adult dairy cow weighing 500 kg and walking one mile/day, expressed in amounts of dry feed needed each day, and dietary concentrations of nutrients in the dry matter.

Level of production	Requirement	Energy concentration of diet (Mcal/kg dry matter)			
		1.8	2.2	2.6	3.0
	Dry matter (kg/day)	6.5	5.1	4.2	3.0
	Concentration in dry matter (%)				
Dry non-pregnant (maintenance)	Digestible crude protein	3.1	3.6	4.1	4.6
	Calcium	0.28	0.36	0.44	0.53
	Phosphorus	0.39	0.50	0.61	0.73
	Magnesium	0.12	0.15	0.18	0.21
	Dry matter (kg/day)	15.9	11.3	8.8	7.5
	Concentration in dry matter (%)				
Producing 10 kg milk/day	Digestible crude protein	5.0	6.5	8.0	9.1
	Calcium	0.29	0.41	0.52	0.61
	Phosphorus	0.27	0.38	0.49	0.57
	Magnesium	0.10	0.14	0.18	0.21
	Dry matter (kg/day)	—	—	20.1[a]	16.4
	Concentration in dry matter (%)				
Producing 30 kg milk/day	Digestible crude protein	—	—	9.0[a]	10.7
	Calcium	—	—	0.50[a]	0.62
	Phosphorus	—	—	0.38[a]	0.48
	Magnesium	—	—	0.14[a]	0.17

3. Energy concentration of the diet
4. Nature and ratio of fatty acids formed from rumen fermentation
5. Physical conditions under which ration is fed.

A demand has arisen in many areas of the United States for a system of feeding dairy cows that permits a single ration to be group fed to all cows in the herd. This demand came as a natural result of the increasing herd sizes, the use of milking parlors and the lack of the type of labor that is needed to give individual attention to cows.

At the Georgia Experiment Station we have been working for some years on the development of what we call "optimum rations." We have defined such rations as follows: "An optimum ration for dairy cows is one designed to provide adequate quantities of ration nutrients balanced to provide the necessary precursors for rumen fermentation which will, in turn, provide to the udder as needed precursors for the production of the inherited volume of milk with its normal content of milk constituents." In our work we have attempted to consider rations as having the following effects:

1. An influence on total intake.
2. An influence on rate and type of rumen fermentation which in turn influences:
   (1) Efficiency of utilization
   (2) Total milk production
   (3) Composition of milk produced.

No claim is made for the uniqueness of this approach and we have borrowed freely from similar research throughout the world.

The ration that we are interested in will be neither a "high grain" nor a "high forage" ration—rather it will utilize the available forages and grains to provide a total ration of such composition that the cows will be well fed under free choice feeding conditions.

The first item in developing the ration was the matter of intake. It was necessary to determine the influence of the digestibility of the ration on intake and the influence of size of the cow and level of milk production on the intake of a ration at a given level of digestibility. Work at several stations has indicated that the detrimental influences of low digestibility of rations on intake is largely overcome when the ration is 65 to 68 per cent digestible. Using rations in this range, we found that 85 per cent of the observed variability in TDN intake was accounted for by the size of the cow, her level of milk production and her gain in body weight. The regression formula which described this relationship was:

TDN intake =
0.28 (milk produced) + 9.5 (body weight) + 2.01 (weight gain).

In practice it was found that the majority of the cows in a herd would consume feed with a relationship close enough to their need to permit the use of single rations. Some cows do not have sufficient appetites or in some cases have too great an appetite to do well on optimum rations. In light of the increasing difficulties in providing individual attention to cows, these cows which do not perform well may simply have to be culled.

Having established the general relationship between body size and milk production and intake of feed dry matter, it was necessary to devise a general ration composition that would meet the several requirements laid down in the original definition of "optimum" rations. The data in figure 5 illustrate the results of many experiments throughout the world on the relationships between measurable characteristics of rations and milk production. Based on this data, the composition of an "optimum" ration is currently being stated as follows:

1. Dry matter digestibility—65% to 68%; 2. Crude fiber content of total ration, 18% to 22%; 3. Crude protein, 12% to 14%; and 4. Calcium, 0.7% and Phosphorus, 0.5%.

In calculating the rations, it is suggested that the grain portion be calculated using calculated net energy values from feeding tables and that the forage portion be calculated on the basis of an actual chemical composition of the forage to be used.

The several feeding standards that have been in use throughout the world have served the dairy industry well. It is frequently argued that they have seldom been used on farms to calculate the individual daily feed allotment for cows. This argument is undoubtedly true, but the general ideas involved in the systems have been widely used in teaching students in schools, in farmer meetings

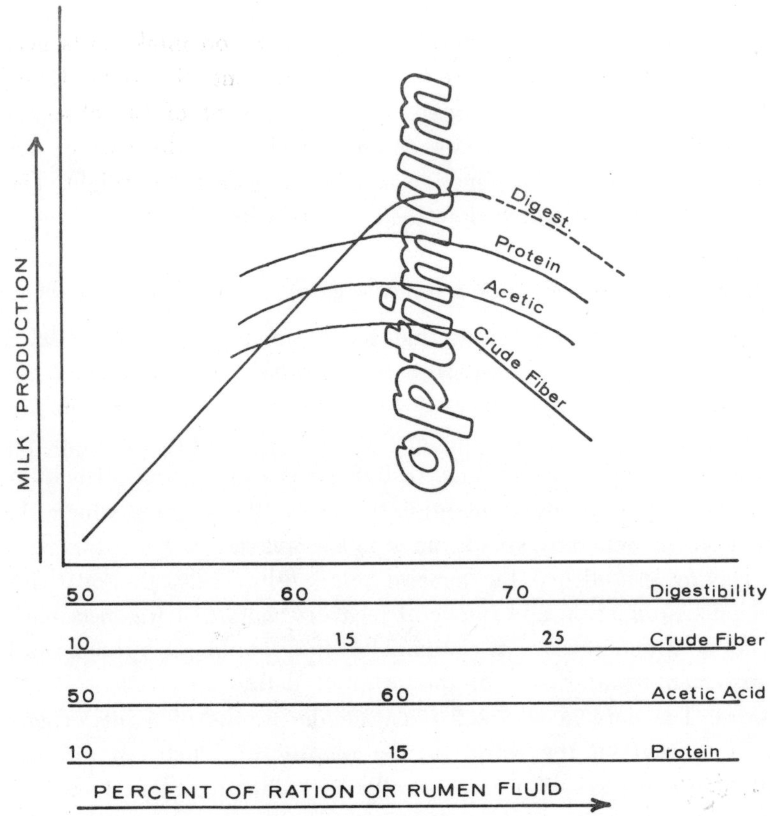

Figure 5 The Relationship between Ration Digestibility, Crude Fiber and Protein Percentage and Acetic Acid Percentage in Rumen Fluid and Milk Production.

and in the popular farm press. The result of all this teaching must be that the systems have greatly influenced the actual farm practice in feeding dairy cows. Each of the systems has merits and difficulties. When these are clearly understood, their usefulness is enhanced.

In science as in all human endeavor, new days call for new ideas. The growing inability of dairy farmers to practice individual cow feeding and care necessitates the development of feeding systems which will permit the cow to meet her individual needs through free choice feeding around a common feed trough. Research and farm experience will further refine the basic information needed to formulate such rations, but the successful use of the current concept of "optimum" rations in both research and under farm conditions indicates that the present plan has merit.

References

1. Blaxter, K. L. Energy Feeding Standards for Dairy Cattle. Nutrition Abstracts and Reviews 20:1. 1949.
2. Blaxter, K. L. Starch Equivalents, Ration Standards and Milk Production. Proc. British Soc. Animal Prod. 1956.
3. Blaxter, K. L. Progress in Assessing the Energy Value of Feedingstuffs for Ruminants. J. Royal Agr. Soc. 123:7. 1962.
4. British Agricultural Research Council. The Nutrient Requirements of Farm Livestock. No. Ruminants. Agricultural Research Council. London. 1965.
5. Conrad, H. R., Pratt, A. D., and Hibbs, J. W. Some Aspects of the Regulation of Feed Intake in Dairy Cows. J. Dairy Sci. 45:685. 1962.
6. Harris, L. E. Symposium on Feeds and Meats Terminology: III. A System for Naming and Describing Feeds, Energy Terminology and the Use of Such Information in Calculating Diets. J. Animal Sci. 22:535. 1963.
7. Heaney, D. P. and Pigden, W. J. Interrelationships and Conversion Factors Between Expressions of the Digestible Energy of Forages. J. Animal Sci. 22:956. 1963.
8. Kielanowski, A. and Zofia, O. Equations for Reciprocal Conversions of Starch Equivalent and Total Digestible Nutrient Values. Canadian J. Animal Sci. 42:211. 1962.
9. Kleiber, Max. Body Size and Metabolism. Hilgardia 6:315. 1932.
10. Kriss, Max. Evaluation of Feeds on the Basis of Net Available Nutrients. J. Animal Sci. 2:63. 1943.
11. Lofgreen, G. P. and Garrett, W. N. Net Energy Requirements, Feed Values for Growing and Finishing Cattle. Feedstuffs, July 22, 1967.
12. Maynard, L. A. Total Digestible Nutrients as a Measure of Feed Energy. J. Nutrition 51:15. 1953.
13. McCullough, M. E. and Smart, W. W. G. Rumen Fermentation and Milk Production. Proc. 1967 Georgia Nutrition Conference.
14. Moe, P. W. and Flatt, W. P. Net Energy Value of Feedstuffs for Lactation. J. Dairy Science. 52:928. 1969.
15. Moore, L. A. Evaluation of Feeds for Ruminants. Feedstuffs, July 16, 1966.
16. Wright, N. C. A Century's Progress in Animal Nutrition. Ayshire Agr. Assoc. Centenary Handbook. 1936.

Chapter 3

Feedstuffs—The Tools of Nutrition

One cannot really differentiate between the role of animals in converting feeds into high quality animal products for human consumption and the role of feeds in supporting animal industries. They are in fact one and the same. With dairy cows, the role is that of a converter of otherwise useless feeds—useless in terms of human food—into milk which is nature's most nearly perfect food. The object of this process in terms of the dairy farmer is profit. Unfortunately, this simple fact is frequently forgotten by many people, sometimes even by the farmer. This forgetfulness occurs when primary attention is given to either the utilization of a favorite feedstuff regardless of its cost or usefulness in dairy rations or when maximum milk production is sought in terms of records regardless of the cost of obtaining those records. Here we will be concerned with feedstuffs as they are and their relative usefulness regardless of their popularity or the beauty of the package in which they come.

Feedstuffs are primarily evaluated on the basis of their ability to provide energy and protein for dairy rations. Vitamins and minerals no longer need be provided by natural feedstuffs. In fact, present non-feed sources of vitamins and minerals provide these nutrients at such low cost that nothing would really be lost if natural feeds were devoid of these nutrients. In balancing dairy rations, emphasis should be given to energy, protein and fiber. Having balanced the ration for these three requirements, the vitamin and mineral contents of the feed used can be determined as a guide to the need for further fortification. The present widespread use of urea as a source of non-protein-nitrogen and the great amount of research dedicated to increasing the total protein that can be replaced from chemical nitrogen sources may well lead to the possibility that feeds will be selected only as an energy and fiber source in the near future.

The dairy cow is primarily a converter of forages and comes well equipped for the job. When economic conditions exist that make the cost of nutrients in grains competitive with those in forages, the cow can adapt to the situation and use large quantities of grain. Under good farming practices, however, this situation occurs less frequently than many people like to believe. Forage, particularly corn silage, is seldom equaled in the production of high quality feed for dairy cows on a per acre basis. The primary problem in forages is the greater management skill required as compared to the production of grain. The possession of this skill is a primary factor in determining which dairy farmer stays in business and which one goes into another occupation.

FORAGES AS FEEDSTUFFS

Concentrated research in forage evaluation as a feedstuff is now nearing its first century mark. The literature is almost unbelievable in scope and equally unbelievable in its lack of uniformity and real usefulness. Within this large volume of results of research, there are many useful discoveries and the overall basic principles of forage utilization have been outlined. In evaluating the results of this research, it may be useful to point out just what good research can and cannot do. Useful research findings result from experimentation designed to outline the basic principles involved in the process under study. In our case, good research into the many problems of producing and using feedstuffs involves studies of the basic principles of production and utilization and measurements on the variation in results that can be expected. Feeding trials comparing one feed to another or one ration to another are of little real value since the normal variation in feeds means that no two trials can be expected to give identical results. Once it has been established that a feedstuff is palatable to cows, its chemical composition and relative digestibility has been determined, there is little to be gained from making comparison after comparison of the feedstuff with other feeds. The idea of associative effects between feedstuffs has been a long held idea in nutrition but one would be hard pressed to come up with a real good example of its importance. In this discussion of forages, therefore, we will attempt to outline the basic principles of forage utilization and will ignore individual forages as such.

The basic principles of forage utilization are outlined in figure 6. Within a given forage, the composition of the particular lot of forage that is available constitutes the basic factor in determining the usefulness of that forage. It will be discussed in detail later, but forage composition is primarily a function of stage of maturity and this is reflected in the relative digestibility of the forage. The data in figure 7 shows the overall relationship between digestibility and animal performance for gain and in figure 8 the same relationship with milk production. The importance of the relative digestibility of the forage

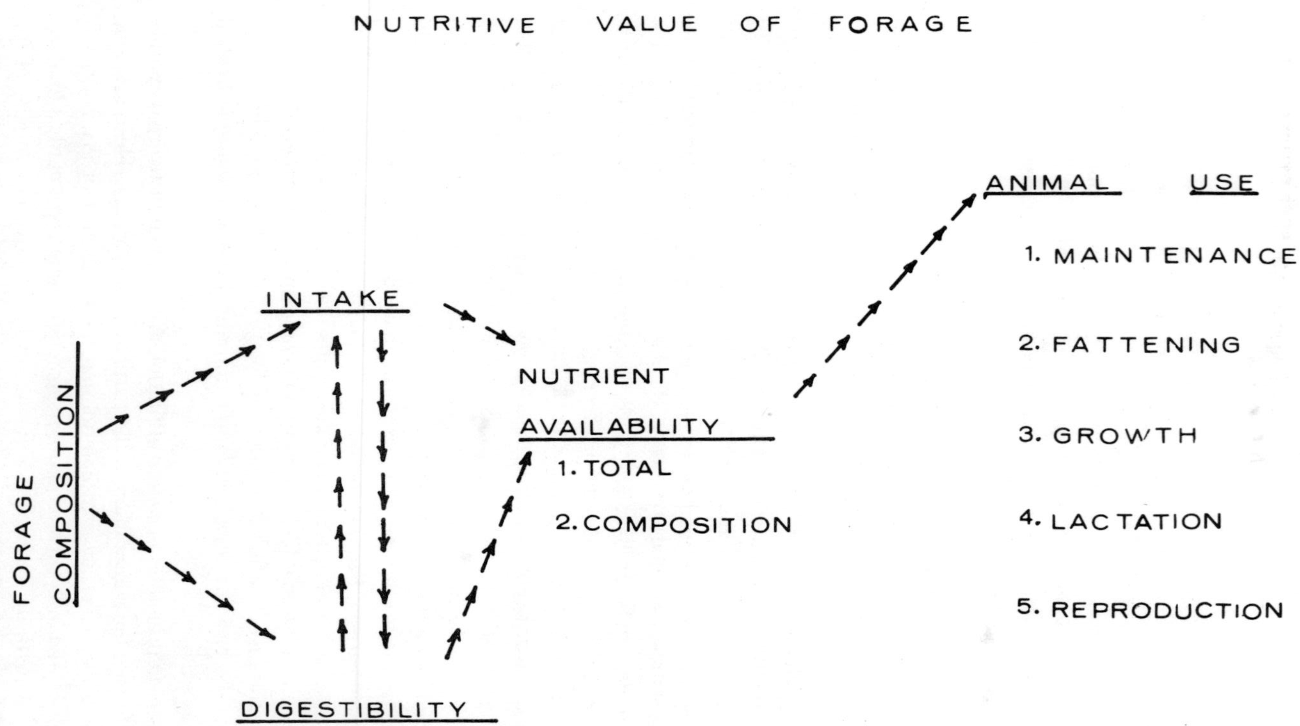

Figure 6 The Several Factors Known to Influence the Efficiency with Which a Forage is Used for Animal Production.

comes from its influence on total intake of the feed and its importance in determining the total quantity of nutrients which are available. This relationship between digestibility and intake cannot be overemphasized. The frequently heard argument of which is more important intake or digestibility is equivalent to the old argument of which comes first the hen or the egg. The important fact is that the two are so interrelated that one comes with the other. Thus to assure both high levels of forage intake and the greatest availability of nutrients, it is necessary to provide dairy cows with forages harvested at the proper stage of maturity to insure a high percentage of the nutrients being digested. Although no hard fast rule can be given, it seems reasonable in modern dairy rations to assume that this level of digestibility should not be below 60 per cent digestibility of the dry matter of the forage. As will be discussed in detail in later chapters, the necessary digestibility (or quality) of the forage to be used will be determined in part by the use for which the forage is being fed. In descending order of requirements for quality of forage for animal use, the general order is fattening, lactation, growth, reproduction, and maintenance.

Since forages usually constitute only part of the total ration, it is necessary

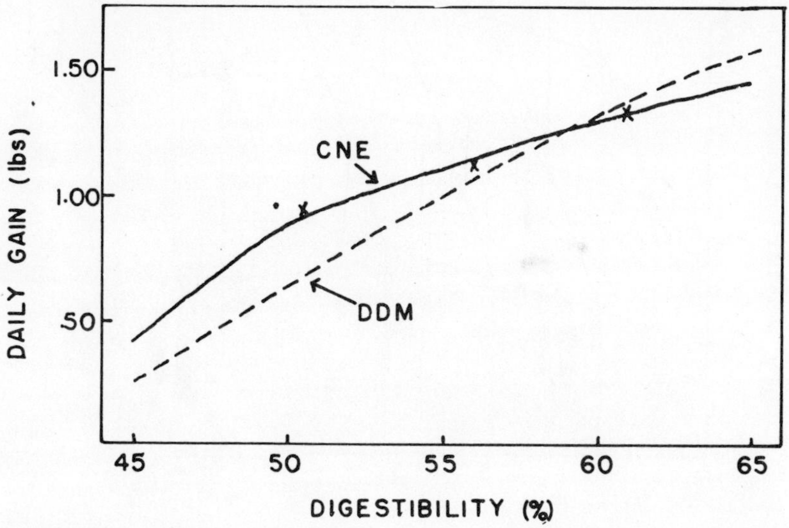

Figure 7 The Relationship Between Dry Matter Digestibility (DDM) and Calculated Net Energy (CNE) and Average Daily Gains of Dairy Heifers Fed Forage Rations.

to consider the relative economics of harvesting forages at less than maximum digestibility in order to harvest a greater quantity of dry matter per acre. This argument has sometimes been carried to great extremes by advocates of both maximum yield per acre and maximum animal production per unit of feed. There is no single answer to the problem since the relative economics varies from year to year and from farm to farm within years. In practice, however, the decline in quality that can be tolerated in the interest of greater

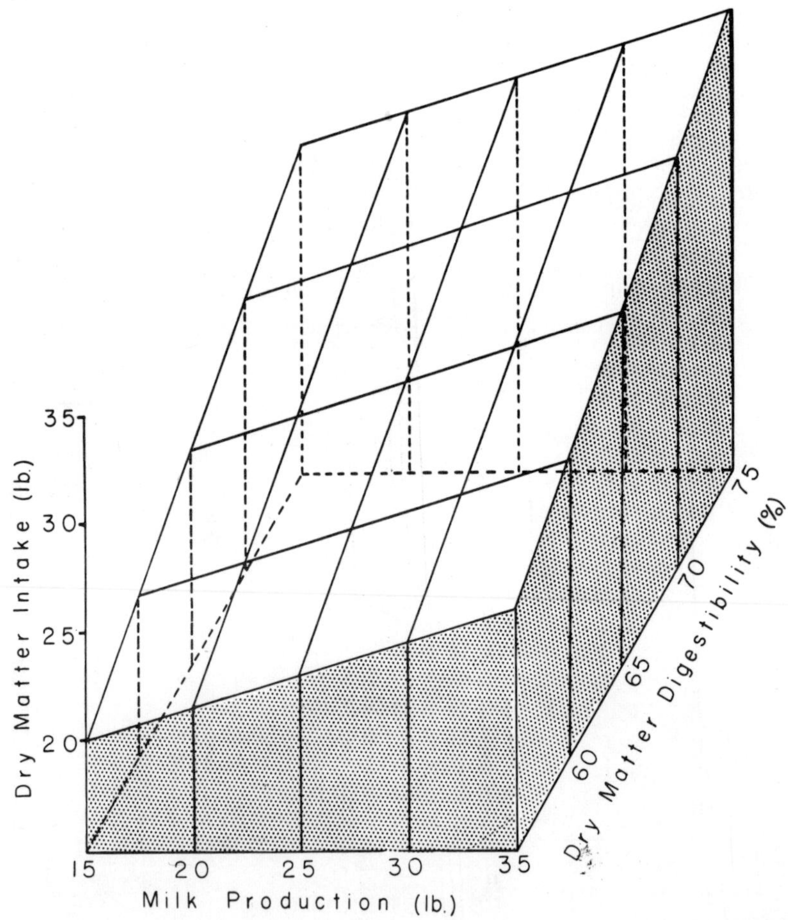

Figure 8 The Relationships Between Dry Matter Digestibility, Dry Matter Intake and Milk Production in Dairy Cows.

production of dry matter per acre is less than many people believe. This is true primarily because of the multiplying effects of declining digestibility on intake. The data in table 4 illustrate the interaction of dry matter digestibility of harvested forage and the ability of that forage to supply nutrients for a dairy herd. It can be seen that delaying harvest of a forage with an initial dry matter digestibility of 67 per cent for a period sufficient to permit a decline in digestibility to 55 per cent reduces the ability of the forage to contribute to the total ration from 75 to 58 per cent for milking cows and from 100 to 57 per cent for heifer feeding. It appears, therefore, that our suggestion that forages be harvested to have a dry matter digestibility above 60 per cent is a good one. The overall relationships between date of harvest, dry matter digestibility, feed intake and milk production are well illustrated in figure 9, which summarizes an outstanding bit of research at the Ohio station by Conrad, Hibbs, Pratt, and Davis.

Table 4 The influence of forage quality on the ability of forage to contribute to total estimated TDN needs for a dairy herd.[a]

Item	Forage Dry Matter Digestibility 55%	62%	67%
Cows			
Total TDN (tons)	204.6	204.6	204.6
TDN from forage (tons)	118.7	141.2	153.4
TDN from grain (tons)	85.9	63.4	51.2
Per cent from forage .	58	69	75
Heifers			
Total TDN (tons)	48.3	48.3	48.3
TDN from forage (tons)	27.5	43.5	48.3
TDN from grain (tons)	20.8	4.8	—
Per cent from forage	57	90	100
Total TDN (tons)	252.9	252.9	252.9
Per cent from forage	58	73	•80
Per cent from grain	42	27	29
Tons Grain Required	141	91	67
Tons Hay Equivalent	293	327	332

[a] 50 dairy cows and 35 replacement heifers over a 12-month period.

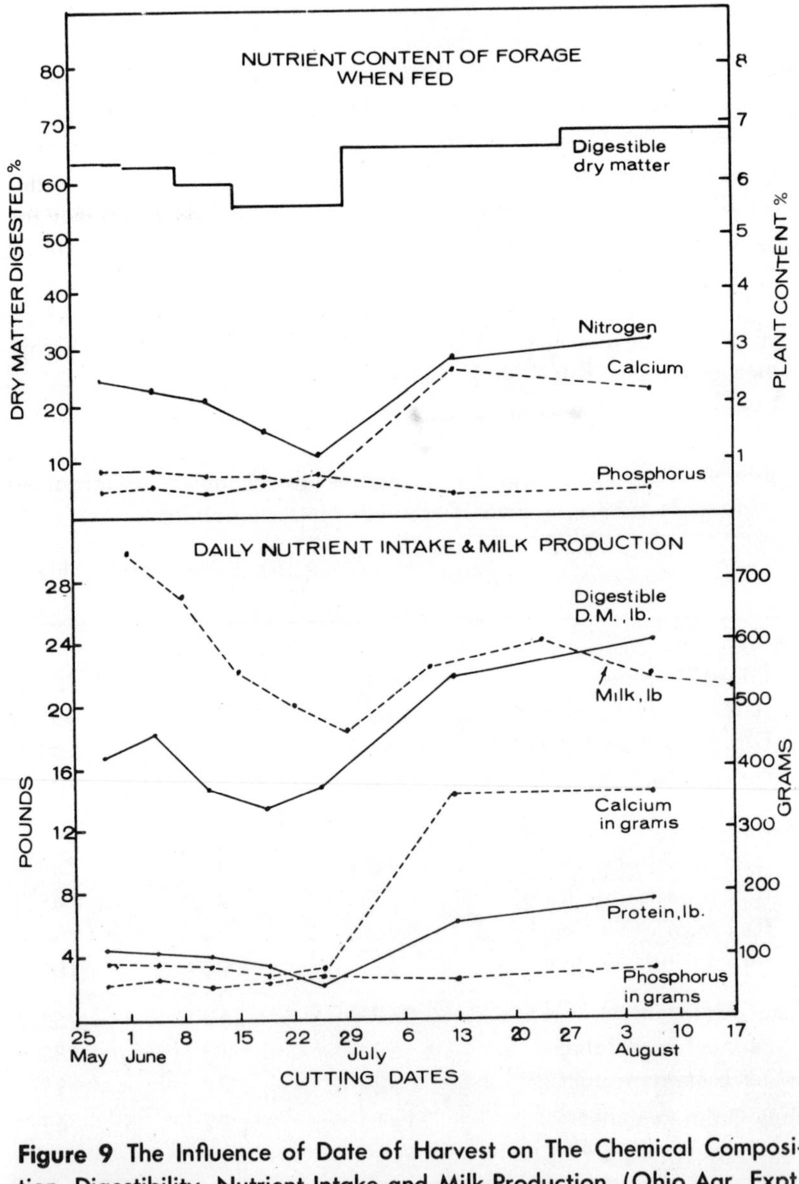

Figure 9 The Influence of Date of Harvest on The Chemical Composition, Digestibility, Nutrient Intake and Milk Production. (Ohio Agr. Expt. Station Bull. 914)

STAGE OF MATURITY FOR HARVEST

It is far easier to discuss harvesting forages to provide 60 per cent dry matter digestibility than it is to do the actual harvesting. Each forage exhibits its own characteristic changes in composition and digestibility with aging and each has its own upper limits of quality. A fact that must be faced is that some forages simply do not lend themselves to feeding dairy cows because their upper limits of quality are either below the accepted quality or they exhibit a level of acceptable digestibility at so young a stage of maturity that they cannot be economically harvested for feed. Only those forages that local research has shown to be useful in feeding dairy cows should be considered for use in dairy rations.

The data in figure 10 illustrate the influence of aging on the percentage of digestible dry matter per unit of forage for several of the more useful crops for dairy rations. Alfalfa is the most widely used legume for dairy feeding and, as can be seen, it should be harvested at not more than an early bloom stage if it is to have an expected dry matter digestibility above 60 per cent. Oat forage declines in digestibility very rapidly after heading and should be harvested when it begins to produce heads above the flag leaf. Timothy is probably typical of many grasses and its digestibility declines almost lineally with aging. Most grasses are best harvested just before heading to insure the desired level of digestibility. Corn silage and probably grain sorghum silage represents a departure from the rule of changing digestibility with age. The corn plant is unusual in that aging is accompanied by a large increase in highly digestible grain which may constitute as much as 50 to 60 per cent of the total dry matter of the harvested material as well as a plant with leaves and stem that is well digested even when mature. The corn plant can obviously be harvested over a much longer period than most forages and decisions on time to harvest corn for silage should include dry matter of the material, quality of the leaves and the prevention of loss of dry matter from degeneration under field conditions.

EVALUATING FORAGES FOR FEEDING

Since forages exhibit so wide a range in chemical composition and feeding value due to stage of maturity at harvest and handling, no general composition from some table of feeds should be used in calculating the feeding value of a lot of forage when calculating modern dairy rations. For this reason, most states and many feed companies have initiated programs for determining the relative feeding value of individual lots of hay or silos of silage. These

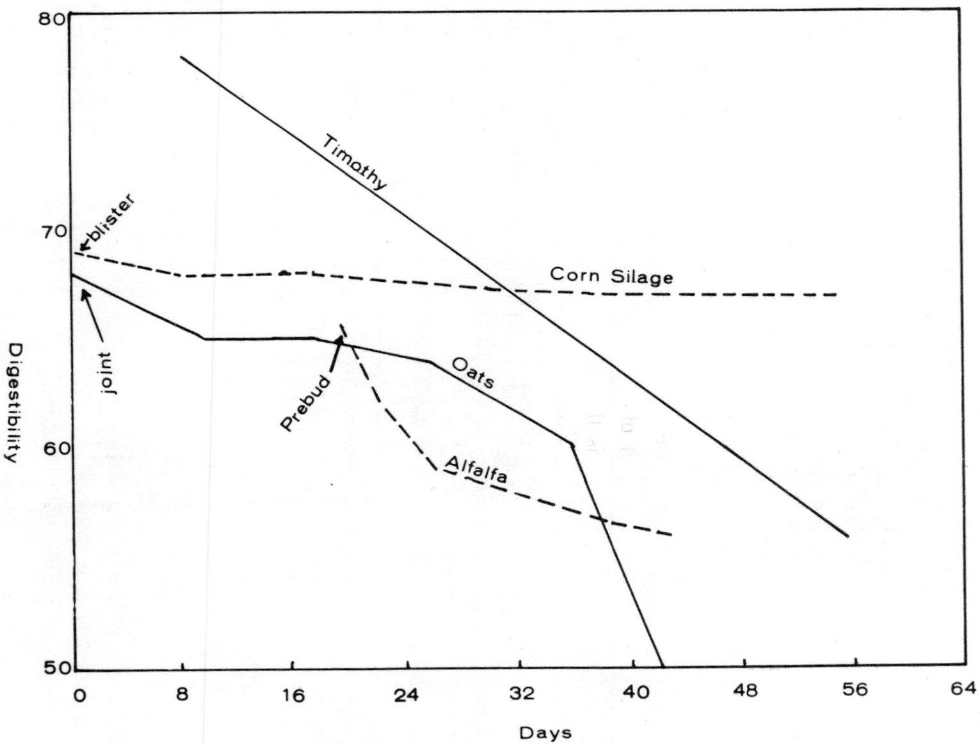

Figure 10 The Influence of Ageing on the Digestibility of Several Forage Crops.

programs have several names—the more common being "forage testing." In such programs, a representative sample of the forage to be evaluated is sent to a laboratory to be analyzed for moisture, crude protein, crude fiber and sometimes calcium and phosphorus. Using this information, the TDN or estimated net energy of the forage is calculated. Many people have questioned the validity of such calculations pointing out the great variation in such calculation. It is true that TDN or net energy calculated in this manner is subject to error, but it is also true that it is far superior to using a table value or to a guess. One point that is frequently overlooked is that the forage is used as only part of the ration and much of the error in calculation is compensated for by other components of the ration. In practice the superiority of this system of determining what value shall be given to the forage portion of the ration has been demonstrated many times. Most states have developed regression formulas for use in their forage testing programs which are more accurate under the conditions prevailing in their area than some universal formulas. Where available, local formulas should be used. When such formulas are not available, good estimations can be made by using the following formulas.

For grass hays:
TDN = 86.2 − 0.68 X where X = per cent crude fiber

For legume hays:
TDN = 62.8 − 0.04 X where X = per cent crude fiber

For corn silage: TDN = 90.93 + 0.39 DP − 1.10 crude fiber

For sorghum silage: TDN = 87.10 − 1.06 crude fiber

For calculating digestible protein for all feeds:
Dig. Protein = −2.79 + 0.88 X where X = crude protein per cent

For converting TDN to estimated net energy for all feeds:
ENE = 1.393 X − 34.63 where X = per cent TDN

HAY AND SILAGE STANDARDS

Despite the fact that millions of tons of hay and occasional silos of silage are sold each year in the United States, there are few lots sold which carry any standard other than some vague statement that it is good or poor hay. In 1965 a committee of specialists in this field was appointed by the American Forage and Grassland Council to establish standards for hay and silage that could be based on forage testing and could permit hay and silage to be rated relative to its value as a feedstuff for livestock. The standards established by the committee and approved by the council are in table 5. Widespread use of these standards could go a long way towards insuring a purchaser of hay that he was getting a feedstuff that was adapted to his needs.

METHODS OF FEEDING FORAGE

Pasture and Green Chop A book such as this written 25 years ago might well have devoted a major portion of the section on feeds to a lengthy discussion of pasture and green chop. Pasture has served the dairy industry well and its contribution cannot be overemphasized. At the moment, pasture

Table 5 AFGC hay and silage standards.

Crop	Standard	TDN	CNE	Crude Protein
Dry matter basis				
Legume hay	Excellent	65% or above	56T or above	18% or above
(Majority of the	Good	58% to 65%	46T to 56T	At least 15%
plant material	Fair	54% to 58%	40T to 46T	At least 12%
originating from	Poor	Below 54%	Below 40T	Below 10%
legumes)				
Grass hay	Excellent	65% or above	56T or above	15% or above
(Majority of the	Good	58% to 65%	46T to 56T	At least 12%
plant material	Fair	54% to 58%	40T to 46T	At least 10%
originating from	Poor	Below 54%	Below 40T	Below 10%
grasses)				
Corn silage	Excellent	65% or above	56T or above	9% or above
	Good	60% to 65%	50T to 56T	9% to 11%
	Fair	—	—	—
	Poor	Below 60%	Below 50T	9% to 11%
Hay crop silage	Excellent	Above 65%	Above 56T	Above 16¢
	Good	60% to 65%	50T to 56T	12% to 16%
	Fair	54% to 59%	40T to 50T	Below 12%
	Poor	Below 54%	Below 40T	Below 12%

Note: To be in a class, a forage must meet both the TDN or CNE and protein requirements of that class.

Hay crop silage is divided into three classes in terms of dry matter content:
1. High Moisture silage—contains less than 25% dry matter.
pH should be below 4.2.
2. Medium Moisture silage—contains 25% to 35% dry matter.
pH should be below 4.5.
3. Low Moisture silage—contains more than 35% dry matter.
pH is no criteria.

constitutes less than 20 per cent of the nutrients used on dairy farms for milk production. This change has come about because of changes in many factors which influence the selection of feed supplies for dairy farms. The primary reasons for de-emphasis of pasture are: (1) the need for making maximum use of land, (2) high levels of production which can be sustained only with a constant supply of high quality feed, and (3) the development of machinery for harvesting, storing and feeding silage and hay which reduced the labor requirements for stored feed.

Numerous experiments have shown that high levels of animal performance can be obtained by letting the animal harvest forage by grazing or by mechanically harvesting fresh forage and feeding it as green chop. When our dairy herds were small in numbers of cows per farm, the necessary investment in machinery for harvesting forage could not be afforded. In recent years, investments in machinery have become a cheaper means of getting work done than investments in labor. Thus harvesting forage has become an even better alternative from an economic point of view. This development has permitted a change from pasture that has long been desired because of the wide fluctuations in feeding value which maximum pasture use provided in dairy feeding programs. Pasture or fields of growing crops used as green chop go through all the stages of maturing typical of the crop being grown and at the same time is constantly subjected to such weather hazards as drought and high temperatures. Feeding programs based on pasture were always feeding programs which were good or bad depending upon weather conditions. When a dairyman is milking 100 cows and has a $100,000 or more investment, he would like to reduce such risks to a minimum.

Pastures will undoubtedly continue to be used on dairy farms for many years particularly for growing heifers and non-producing cows where variations in feed quality are of somewhat less importance than with high producing dairy cows. It is also true that many dairy farms have land that is not suited to intensive crop production and pasture provides the best use for such land. So long as this type of land is used for dairy farming, pasture will be needed for such herds. The constant decline in percentage of nutrients obtained from pasture by dairy herds in the United States is ample evidence of the declining importance of pasture in feeding dairy cows.

When pasture is used, its feeding value is determined by the digestibility of the dry matter and the quantity of forage available. Thus young pasture can be of exceptional high feeding value and the level of intake can be very high when high digestibility is accompanied by adequate forage. The forage available is subject to rapid changes in digestibility under hot dry conditions.

Many trials have shown declines in digestibility of one percentage point per day. In contrast to this rapid change, forage grown under conditions of short day length, cool temperatures, and adequate rainfall which retard maturity can remain highly digestible for long periods of time. Since the composition of the forage and its digestibility cannot be measured from day to day, pasture forage cannot be handled in today's modern feed programming techniques. This factor, along with those listed above, leaves pasture as the least desirable of the available forms of forage feeding for modern dairy feeding programs.

Hay and Dehydrated Forages The basic principle in hay making or forage dehydration is to harvest forage at a specified stage of maturity, preserve the nutrients by drying and store the nutrients in a manner that will not result in further deterioration until fed. The primary problems in the process are those posed by weather conditions when field drying is practiced and the large amount of storage space required because of the bulky nature of the hay. The detrimental effects of rain during the curing process are dependent upon the time of the rainfall and the delay caused by rain. A small amount of rain immediately after the hay has been cut is of little significance while even a small amount of rain once the forage has dried can be highly detrimental. It is this problem in curing forage under adverse weather conditions that has caused a steady decline in hay production on dairy farms. The other factor is of course the problems of handling, storing, and feeding a bulky feed.

In an attempt to overcome the two major problems of hay making, considerable research has been done on dehydrating and pelleting forages as an alternative to field curing and baling. There is little doubt that dehydrating and pelleting is a satisfactory answer to the problem, but to date, it has not proven economically feasible.

Since a large portion of the nation's hay crop moves through commercial channels, there has been considerable interest in pelleting or cubing to reduce the bulk of hay and reduce costs of transportation. To date, this has not become a major factor in hay marketing, but it obviously has merit. Another use of pelleting that was widely touted some years ago was to increase feed intake. The major response in intake was with low quality forages and the process has yet to make significant gains in feeding programs since low quality forages are not suited to milk production.

As with pasture, the value of hay in feeding programs is dependent upon its digestibility and intake which in turn is determined by the crop grown and the stage of maturity at harvest. There is no doubt that hay is a perfectly good method of harvesting and storing forage for feeding dairy cows. The decline in its use is due to factors other than its feeding value. The future of hay in dairy rations will undoubtedly be determined in large part by new developments in methods of harvesting, storing and feeding that reduce weather losses and labor requirements.

Silage The need for conserving feed at a period of lush growth for feeding in periods of deficient growth is a problem as old as agriculture. Early peoples stored grains in pits and probably wondered why the same system could not be used to store forages. Thus it is not surprising that the first known attempts at making silage depended upon burying fresh cut forage in the ground.

With the advent of modern farming methods, silage has lost many of its "backbreaking" labor characteristics and enjoys a role in modern farming which is unprecedented. Despite the current importance of silage as a method of conserving feed, it is the least understood and most abused of all systems of forage farming. It is our purpose here to explore the events which occur in the silo, the factors which influence the silage making process, and to outline some general principles of good silage making.

WHAT IS SILAGE?

Traditionally, silage has been defined as the end result of the preservation of a crop under conditions suitable for the development of sufficient lactic acid to maintain the mass in a constant state of preservation. In brief, the crop is stored under anaerobic conditions, supplied with available carbohydrates and maintained at the optimum temperature for lactic acid producing organisms to produce sufficient acidity to prevent further decomposition.

The traditional method of silage making has been subjected to many alterations. In northern Europe there developed a process of silage making in which the crop was sterilized by inducing severe heating. The forage was piled in loose stacks and permitted to heat to 145°–160° F. before it was packed. This method prevented further bacterial breakdown but resulted in severe loss of nutrients and a drastic reduction in the digestibility of those remaining.

The opposite approach is that of maintaining the temperature low enough to prevent bacterial growth. In theory this is an ideal method for forage preservation. In practice it is frequently unsuccessful. The method requires the use of high moisture materials, heavy packing and rapid silo filling. A further factor in the ability of micro-organisms to adapt to sub-optimum conditions or in some cases for wide-range organisms to develop. Although this method sometimes works, it is doubtful whether it is worth the risk.

With the advent of research in silage fermentation, a logical alteration was proposed in the form of adding chemicals, chiefly mineral acids, to increase acidity immediately and thus minimize bacterial fermentation. This concept has received wide attention and is used throughout the world. Although basically sound in concept, the economics of its use in practical silage making are still a much debated subject.

Regardless of the method of preservation used, workers in agriculture have long been impressed by the unexplainable variations observed in silages made from similar crops under apparently similar conditions. A typical example is presented in table 6 which summarizes some work at the U.S.D.A. Station at Beltsville by Gordon. The silages were made from a first cutting of an orchard grass-ladino clover mixture harvested without wilting and simultaneously stored in upright silos. The only apparent differences in the forages was a higher sugar content in 1954. The addition of a preservative

Table 6 Variations in composition and feeding value of silage made from similar forage.

Item	1953		1954	
	Untreated	Kylage	Untreated	Kylage
Composition as Stored:				
Dry matter (%)	19.4	20.6	18.4	18.3
Crude Protein (%)	18.0	17.2	16.7	16.8
Sugar (%)	5.4	5.6	8.8	9.1
Composition of Silage:				
Dry matter (%)	22.9	23.9	23.9	24.6
Crude Protein (%)	19.2	19.1	16.0	16.0
Sugar (%)	0.2	0.3	0.5	0.8
pH	4.98	4.52	3.76	3.67
Acetic acid (%)	3.25	3.05	1.98	1.51
Butyric acid (%)	3.05	1.05	.03	.03
Lactic acid (%)	1.80	4.90	7.78	8.00
Results of Feeding Trials:				
Dry matter from silage consumed/cow/day (lb.)	15.2	17.7	21.9	21.5
30-day decline in milk production (%)	12.9	11.8	2.6	2.1

in 1953 resulted in an improvement in an already excellent silage. The preservative was without effect on feeding value of the silage in both years. Attention should be directed to the fact that the silage made in 1953 apparently greatly lowered the feeding value of a good forage.

The example cited above reflects the many problems inherent in silage making and points out the need for a clearer understanding of the silage making process.

WHAT HAPPENS IN THE SILO

The chemistry of silage cannot be separated from the microbiology of the silo since silage fermentation is a bacteriological process. The chemical changes in the silo will be considered in terms of silage made without the addition of a preservative. With this as a background, the specific effects of adding various forms of preservatives will be considered separately.

Phases of Silage Fermentation Barnett divided silage fermentation into five phases and this division will form the basis of discussion. In figure 11 these phases have been outlined along with general examples of several factors commonly considered in silage making. Phases 1, 2 and 3 are completed during the first three to five days and are never clearly defined. The rapidity of this part of silage making should not detract from its importance. The real basis for successful silage making is laid here. Phase 1 begins with the silage being placed in the silo. This phase includes the final respiration of plant cells with the production of heat and carbon dioxide. This phase has great bearing on the future of the silage since the heat produced largely determines whether or not an optimum temperature will be provided and the carbon dioxide produced reduces the air supply and provides anaerobic conditions for lactic acid bacteria. Phase 2 is important because it marks the end of plant cell respiration and the beginning of acid production. This production of primarily acetic acid does much to lower the pH sufficiently to prevent the growth of those bacteria which otherwise might produce undesirable compounds in the silage. Since the acetic acid producing bacteria cannot tolerate a low pH, their numbers decrease rapidly and the lactic acid forming bacteria begin to increase. Thus phase 2 develops into phase 3 with the initiation of lactic acid formation. The first few days also includes the settling of the forage in the silo and the relative rate of seepage increases rapidly, reaching its peak about the fourth or fifth day.

Phase 4 usually begins three to five days after ensiling and requires 15 to 20 days for completion. This is the phase that determines the success of the attempt at silage making. The lactic acid bacteria dominate the flora

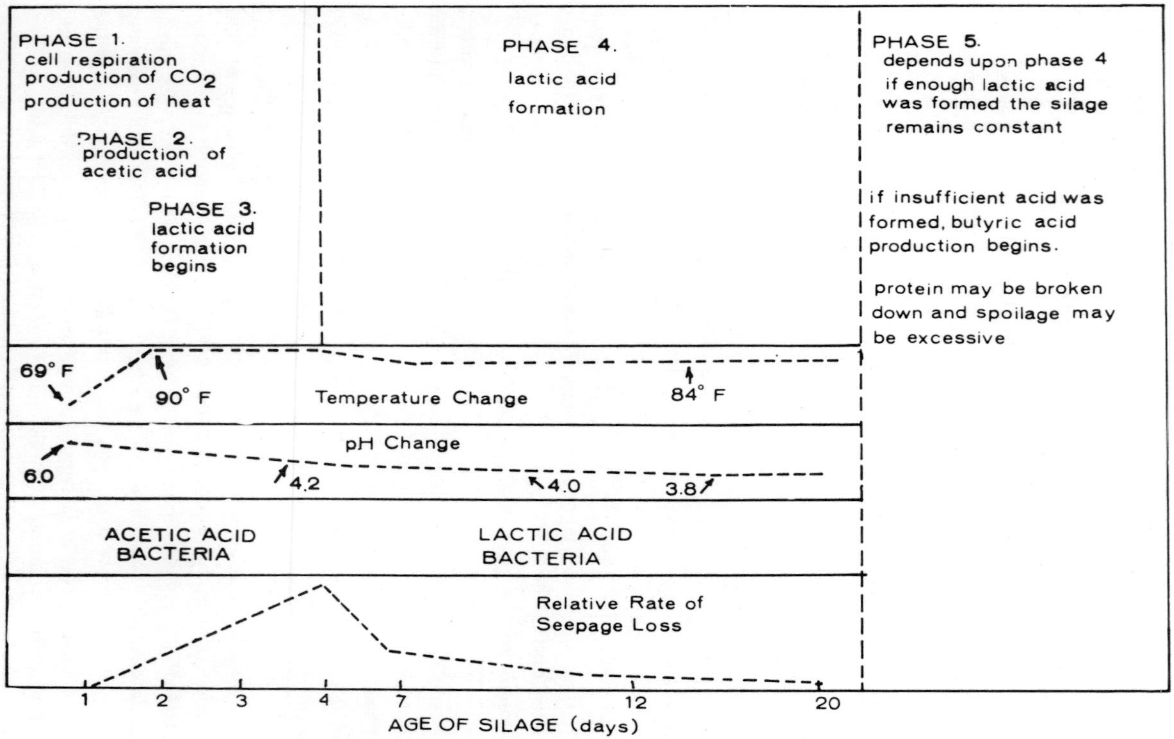

Figure 11 A Schematic Representation of the Five Phases in Silage Fermentation.

and gradually increase the lactic acid content of the silage until the acidity becomes great enough to stop further bacterial action.

Phase 5 represents an indefinite period which reflects the relative value of the changes which occurred in the first four phases. If sufficient lactic and acetic acids were formed to prevent further bacterial action, phase 5 is merely a period during which the silage remains constant. Insufficient acid production during the first four phases, however, leaves the silage subject to decomposition by non-desirable bacteria such as those which produce butyric acid. When this occurs, the normal silage chemistry is unpredictable. Bacteria capable of producing butyric acid will use not only carbohydrates but also lactic acid as a medium. The bacteria also frequently attack the amino acids and protein of the silage which results in the formation of ammonia and volatile fatty acids. This type of process has the further disadvantage of maintaining or increasing the pH of the mass which tends to permit the fermentation to continue until the available energy supply is exhausted.

The five phases discussed here are affected by silage temperature, dry matter content of the ensiled crop, the pH of the silage and the quantity of needed carbohydrates available in the crop. Since each of these factors is to some extent controllable, each should be considered in some detail.

Effect of Temperature As noted earlier, temperature is one of the factors affecting the success of silage making since the lactic acid bacteria have an optimum temperature for growth. In figure 12, the effects of silage temperature on pH and total acidity are shown based on the work of Shaw, *et al.* at the Washington station. This work substantiates the observations of other workers that the optimum silage temperature ranges from 80° to 100° F. In practice, some idea of the relative temperature can be obtained from the appearance of the resulting silage. Three distinct types of silage can be accredited largely to silage temperature. *Seriously under-heated* silage is usually a drab green in color, has a strong odor, the soft tissues are slimy, and an insipid taste with a pH of 5 or above is present. A *properly heated* silage is light green to yellow in color, has a pleasant vinegary odor, tissues are firm and it has a sharp acid taste indicative of a pH below 4.5 per cent. *Overheated* silages have characteristics which are influenced by the degree of heating. The color of such silages ranges from brown to black and odor from slightly burnt sugar to charred hay. Items such as pH and taste are unreliable guides since it is difficult to determine the exact degree of fermentation which occurred.

As noted earlier, silage temperature is largely due to cell respiration.

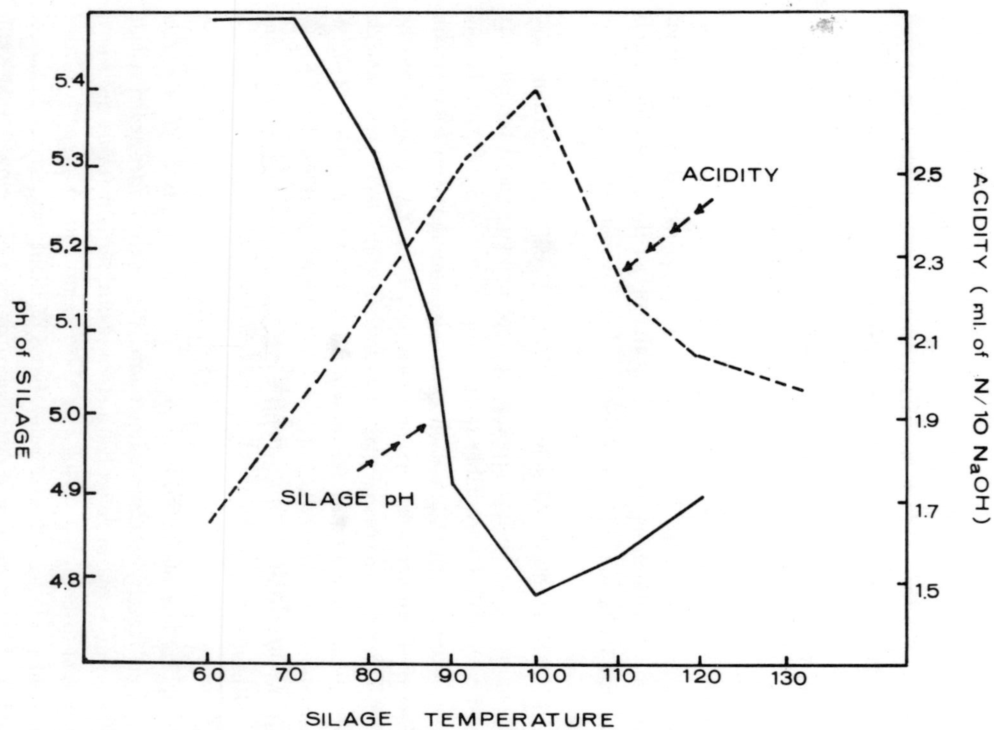

Figure 12 The Influence of Silage Temperature During Fermentation on the Development of Acid and the Reduction

Table 7 Early changes in silo temperatures as affected by temperature of ensiling. (Georgia data)

Item	Temperature					
	Of forage as ensiled	Of silage material days after ensiling				
		1	2	3	4	5
Ensiled on hot day—						
Temperature	90	85	82	81	78	75
higher than final temperature	105	90	89	85	85	85
Ensiled on cool day—						
Temperature	62	67	70	71	72	74
lower than final temperature	69	91	90	90	85	85

As such, the final fermentation temperature is largely determined by rate and method of silo filling. The data shown in table 7 seem to indicate that the temperature at which forage is added to the silo does not control the final silage temperature. Thus, forage ensiled on a hot day may have a temperature greater than desired but it cools down to a temperature determined largely by cell respiration. Conversely, silage ensiled on a relatively cool day increases in temperature with cell respiration. Table 8 presents data on the relative change in silo temperature throughout the day. The small diurnal variation in temperature adds emphasis to the necessity for proper control of silage temperature during the ensiling process. Since temperature is dependent upon the amount of air or oxygen available for bacteria and

Table 8 The diurnal change in temperature in silos. (Georgia data)

Silo	Temperature °F											
	8 am	10	12	2 pm	4	6	8	10	12	2 am	4	6
1	70	71	71	72	74	74	74	73	72	72	71	71
2	74	78	79	78	79	78	76	76	75	75	74	74
4	72	72	72	73	74	74	74	73	72	72	72	72
5	86	85	85	84	84	84	84	83	83	82	82	83

for cell respiration, silage temperature is largely a function of degree of packing, length of cut of the forage and moisture content. Wet forage may require a heating period prior to packing while dry forage may need extra packing to prevent overheating.

Effect of Dry Matter Content of Forage The dry matter content of a crop probably exerts no direct force on the events transpiring in the silo. Its importance stems for its usefulness as a relative measure of such items as stage of maturity, protein content and the relative difficulty of packing the silage. In figure 13, the relationships between the dry matter content of the ensiled forage, the resulting pH of the silage and the loss of seepage are shown. The data for pH were adapted from the work of Axellson and for seepage loss from Murdock. (It should be emphasized that the data presented in tables throughout this publication cannot be used to calculate values for other silos.) The effect of a low percentage of dry matter probably stems from low temperatures resulting from too rapid compaction of the mass. The high pH resulting from high percentages of dry matter may well be due to the inability to control temperature and the inability to provide an anaerobic condition in the silo.

Seepage loss follows the logical pattern since the flow would be expected to decrease with increasing dry matter content of the forage.

The data would seem to indicate that silage is most easily preserved when the dry matter falls in the range of 25 to 35 per cent.

The Effect of pH The pH of silage is a widely used measure of the success of the ensiling process. Most workers would agree that its use is justified only if properly interpreted and accompanied by other information. Of primary interest is the apparent lack of perfect correlation between the pH content and a specific acid constituent. Since silage acids are mixtures, it would be expected that their relative effect on pH would depend upon the degree of ionization and the buffer capacity of the silage mass. In general, however, as the pH decreases it reflects an increase in lactic acid, total acidity and amino acids while denoting a decrease in the volatile acids.

The buffer capacity of the ensiled crop is of considerable interest in silage making. In general, plants such as alfalfa and soybeans are more difficult to ensile than the grasses. When the high ash and protein content of legumes are taken into account, this difference would be expected. In table 9 the relative buffer capacity of four common crops is shown. The data are from the classic work of Wilson. In the case of legumes, only the highly ionized sulfuric acid was capable of lowering the pH of the crop to any marked degree. The lactic and acetic acid which normally preserve silage were

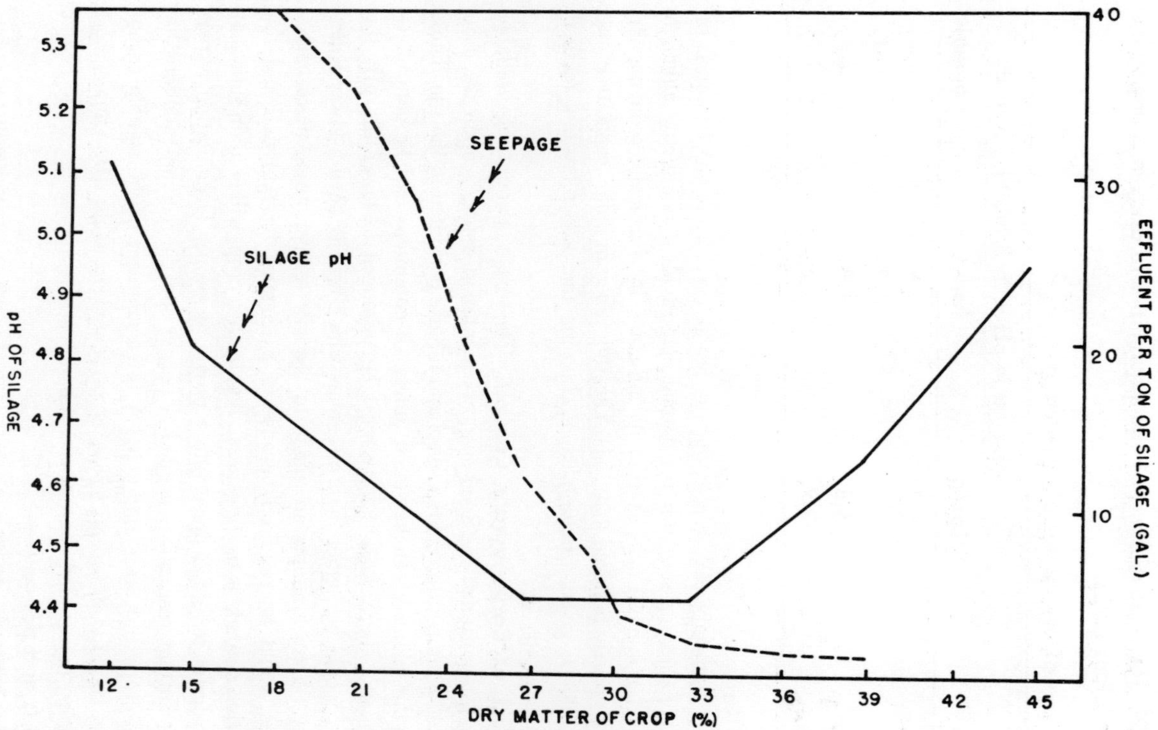

Figure 13 The Influence of the Dry Matter Content of the Crop on the Change in pH and the Amount of Seepage During Silage Fermentation.

Table 9 The relative effect of equal quantities of acids in changing the pH of various crops.

(One gram of dry matter of each crop).

Crop	Change in pH with 3 ml. N/10 acid		
	Lactic	Acetic	Sulfuric
Alfalfa	4.7	4.8	4.4
Corn	4.0	4.3	3.4
Soybeans	4.8	4.9	4.7
Sorghum	3.9	4.3	3.2

relatively ineffective in overcoming the buffer capacity of the legumes. In the case of corn and sorghum, the normal silage acids were effective although the stronger mineral acid exerted its influence by drastically reducing the pH of the poorly buffered materials. This difference in buffer capacity is one that should be given attention when choosing the proper ensiling method for various crops.

Effect of Carbohydrate Content The exact amount of available carbohydrates needed for good silage fermentation depends upon the crop ensiled, the temperature of the silage and the use permitted in the silo. Thus, no specific quantity can be stated unless all conditions can be controlled. In legumes, where more acidity must be produced to lower the pH, the quantity would be higher than in cereals which usually have a lower buffer capacity. When attempts are made to use low temperatures, more carbohydrates would be required because of a sluggish fermentation and the production of weak acids. In general, non-legume crops will contain sufficient carbohydrates for good preservation providing proper fermentation is permitted. Further research is needed to permit specific recommendations for specific crops and crop conditions. A brief summary of the soluble carbohydrate content of several crops is presented in table 10. Under most conditions, the needed content of available carbohydrates appears to center around 6 to 8 per cent.

Why Silage? Today's livestock farmer operates a manufacturing enterprise. The plant consists of his dairy cows or beef herd, his raw materials the forage and grain he feeds, his finished product the milk or meat he sells and, his purpose—profit. We are concerned here with the raw materials used in the manufacturing process. At the moment, the farmer has three alternative sources of feed: (1) pasture or green chop, (2) dried feed in the form of hay, stored or dried grains, and (3) wet feeds in the form of silage. If one is feeding a dozen milking cows or a few steers, economics may dictate the use of pasture. Today's large herds and feeding operations require the provision of such large quantities of feed of uniform quality that stored feeding is the alternative of choice. Drying crops for storage as hay or grain is a perfectly good system of handling feed. For many years it was the method of choice on most farms. Why then the current shift to silage and wet grain? The answer lies in three basic advantages for ensiling: (1) the reduction of losses in harvesting and curing (losses due primarily to weather), (2) the ability to handle the crop from field to feeder by mechanization, and (3) the possibility of using fermentation to make a feed better suited to the needs of the livestock to be fed than the original crop. These advantages of silage should be kept constantly in mind since many of today's recommendations for silage represent a compromise between the advantages listed and the hazards of hay making. Thus wilting silage reduces the advantage of reducing

Table 10 The water soluble carbohydrate content of several crops used for silage.

Crop	Stage of maturity	Per cent water	Water Soluble Carbohydrates	
			Dry basis	Wet basis
		(%)	(%)	(%)
Alfalfa	Beginning to bloom	74.5	4.26	1.09
Barley	Heads just showing	80.7	7.99	1.54
Clover, red	In bloom	75.9	7.52	1.38
Perennial rye grass	In bloom	76.6	10.21	2.39
Soybeans	Pods formed	78.2	2.13	0.46
Corn	Milk stage	77.6	20.25	4.54
Oats	Heading	76.1	9.93	2.37
Sudan grass	Seed—milk stage	70.4	13.41	3.97
Cowpea	Beginning to bloom	79.0	6.55	1.38

weather hazards and a decision to wilt must be made on the basis that the values to be gained are greater than the risks involved.

Since silage and hay are alternative methods for storing forage crops, it might be useful to consider the relative values of the two processes. One can find in the literature of research examples of individual experiments to prove either process good or bad. Proper use of research data dictates the use of average results covering many weather and production situations. The data in table 11 represent the average results from many experiments in several states. We have a natural tendency to ignore what we cannot see. Thus we tend to feel that a crop of hay that is harvested and stored without rain represents a perfect conservation of the crop. The research data in the table indicates that even with barn drying, field losses average 15 per cent and storage losses 6 per cent for a total loss of dry matter with barn-dried hay of 21 per cent. This loss is very near the 25 per cent average dry matter loss which accompanies the storage of high moisture direct-cut forage in tower silos. No one wants the odor and seepage problems which go with direct cut forage, but at its height of popularity in the 1940's, it resulted in no greater average loss of dry matter than average hay making methods. The results shown in the table indicate that maximum efficiency in storing forage as silage is reached with a crop which contains 60–70 per cent moisture. Such a crop has a field to cow loss of some 17 per cent dry matter but, through fermentation, an 8 per cent increase in feed efficiency which provides about 14 per cent more milk per acre than barn-dried hay. Further decreases in dry matter content leads to greater field losses and the lack of fermentation to declines in efficiency of feed utilization which "wipe-out" any additional gain due to reducing storage losses.

PRINCIPLES OF MAKING AND FEEDING QUALITY SILAGE

1. Quality silage begins in the seed store For decades, many people have clung to some vague idea that by processing plant materials through a silo one can make low quality feeds into high quality feeds. This idea has sometimes been kept alive by silo salesmen and others who have a "monetary ax to grind." In feeding both dairy and beef animals, a good rule to follow in selecting crops for silage is this: "if it will make milk or beef without ensiling, it will make milk or beef as silage." Corn silage is the major silage crop in the United States for one very good reason: corn is the greatest yielder of high quality forage available to most American farmers. The casual observer of the agricultural scene must be impressed with the constant procession of "wonder" crops which appear in all the vivid

Table 11 Relative value of hay crop silages compared to barn-dried hay for total milk yield per acre.[a]

Forage: Moisture content and type of storage	Potential DM in field	Av. field and harvesting losses	Net tons DM stored	Av. storage loss	Net DM for feeding	Efficiency	Equivalent tons of hay DM	Total value relative to hay
	tons	%	tons	%	tons	%	tons	%
70% moisture and higher								
—Conventional tower silo	100	2	98	23	75	110	83	104
60 to 70% moisture								
—Conventional tower silo	100	5	95	12	84	108	91	114
35 to 60% moisture								
55% moisture								
—Conventional tower silo	100	8	92	12	81	101	82	103
50% moisture								
—Air-tight silo	100	10	90	3	87	101	88	110
40% moisture								
—Air-tight silo	100	13	87	2	85	101	86	108
Barn-dried hay								
(25–35% moisture)	100	15	85	6	80	100	80	100

[a] Adapted from Animal Husbandry Mimeo. Series No. 3, Cornell University.

colors of modern photography to play a short-time role, only to be succeeded by the next "wonder" crop. This panorama of miracle crops attests primarily to the ability of American advertising since the statistics of crop production indicate the continuous use of corn, sorghum, alfalfa and other high quality crops by the more successful livestock farmers.

2. Harvest at optimum stage Farmers who plant crops which are capable of yeilding high quality feed always produce high quality feed. They do not always harvest it. The point is that, given normal growing conditions, a crop of alfalfa goes through a stage when it can be made into high quality silage. This stage, late bud to early bloom, has been missed if silage is being harvested at full bloom. The fact that the resulting silage is poor has nothing to do with the crop. The farmer simply permitted a good crop to go by. *No uniform, nationwide recommendations can be made since crops respond differently in each climatic condition.* This point cannot be over-emphasized. Farmers in each state should follow the recommendations for their state no matter how good the results look from some other area.

3. The harvesting operation establishes proper conditions for ensiling the crop In addition to moving the crop from field to silo, the harvesting operation can be utilized to do two things: (1) adjust the moisture content of the crop by wilting and (2) improve the compaction within the silo by adjusting the length of cut. Neither of the points require great discussion although in practice they have been the subject of prolonged debate. The optimum dry matter content for silage is 30 per cent to 40 per cent and there is no consistent data to indicate any improvement in the production of milk or meat at greater dry matter contents. The length of cut should be aimed at producing a maximum of the forage being cut ½-inch or less. For dairy cows, a length of cut shorter than this frequently results in butter fat difficulties.

4. The silo must give protection from air and water Since the silo represents a major purchase in silage-feeding programs, it is easy to overestimate the role of the silo in producing quality silage. Actually, a silo is equally capable of housing good and poor silage. The silo has the dual role of protecting the quality of the silage stored in it and expediting mechanical silage removal and feeding. Under good management, no silo performs this function better than the conventional upright silo. The key is good silo management. The research of Browning at Mississippi clearly indicates that storage losses of silage in a well constructed upright silo are no greater than in other silos when they are properly filled and securely capped.

5. Rapid filling and even distribution insures good control of

air Since air is the chief enemy of good silage, this point cannot be given too much consideration. The primary damage from air taken into the silo with the crop occurs in the first 3 or 4 days. The more rapidly a silo can be filled and the more thoroughly it can pack the less air damage can result. Again, no lengthy discussion is required. This is a positive rule of making good silage and, if violated, can result in poor silage from good forage.

6. Additives can be used to improve or alter fermentation In the last 75 years, additives have been the primary subject of research and propaganda in silage production and utilization. Inherent in all the discussions is the basic question of the value of fermentation in the silo. When forage with enough moisture is placed in a silo, the bacteria convert carbohydrates to acetic and lactic acid until the pH reaches 4.0 or below when fermentation stops. Once the acidity has reached this point, the silage will keep for years providing it is protected from air and water. The nature of the fermentation is determined by moisture content, packing in the silo and the quantity and type of carbohydrates available for fermentation. The idea of controlling this fermentation has resulted in continual research on additives for silage. When high moisture-high protein forages are ensiled, some help from additives may be useful. With corn, sorghum or wilted crops additives are of doubtful value.

During the last 75 years, the following ideas have appeared at least twice. The trend is for an idea to show up with some research to support it; some farmers testify that it solved all their problems and greatly improved their silage; the practice disappears only to come back 20–30 years later.

1. Additives Common salt, carbon dioxide, formaldehyde, carbon bisulphide, sulfur dioxide, sodium metabisulphite, benzoic acid, borax, salicylic acid, citric acid, cream of tarter, hydrochloric acid, formic acid, bacterial cultures, enzymes, whey, molasses, potato flakes, beet pulp, citrus pulp, and corn.

2. Mechanical control Direct steam heating of the silage, heating the silage with an electrical current, vacuum silage and weighting of the silage with stones.

In 1969, the list of additives that have merit is very small. The addition of any feedstuff to silage usually has merit providing the cost is reasonable simply because the feed added increases the feeding value of the total silage. For both beef and dairy cattle, corn and sorghum silages can be used as carriers for urea. By adding 10 pounds of urea per ton of forage being ensiled, the urea can be fed without feed refusals by cattle. The urea does not improve the quality of the silage, the silage only serves as a convenient carrier. For fattening, adding 20 pounds of ground limestone may improve feed utilization

Table 12 Feed ingredients—Their composition and energy value.

Ingredient	Crude protein	Crude fiber	Calcium	Phos-phorus	Mag-nesium	Potas-sium
	%	%	%	%	%	%
Alfalfa Meal, DEHY	17.0	25.0	1.30	0.23	0.29	2.49
Animal Fat	-	-	-	-	-	-
Barley	11.5	7.0	0.06	0.36	0.12	0.56
Beet Pulp, dried	09.0	19.0	0.60	0.10	0.27	0.16
Brewers Grains, dried	25.9	15.0	0.27	0.50	0.19	0.06
Citrus Pulp, dried	06.6	13.0	1.90	0.12	0.16	0.62
Corn	09.0	02.0	0.02	0.29	0.15	0.33
Corn and Cob Meal	08.1	08.0	0.04	0.27	0.15	0.53
Corn Cobs	02.3	35.0	0.12	0.04	0.06	0.76
Corn Gluten Feed	22.0	09.0	0.46	0.77	0.29	0.60
Cottonseed Meal (Expeller)	41.0	16.0	0.15	1.00	0.54	1.20
Cottonseed Meal (Solvent)	42.0	12.0	0.15	1.10	0.58	1.47
Cottonseed Hulls	04.0	43.0	0.14	0.09	0.13	0.78
Distillers Dried Grains	27.0	12.0	0.09	0.37	0.06	0.09
Distillers Dried Grains/Solubles	27.0	08.5	0.12	0.68	0.06	0.09
Distillers Dried Solubles	27.0	04.0	0.35	1.30	0.64	1.74
Hominy Feed, Yellow	10.5	05.0	0.05	0.53	0.24	0.67
Linseed Meal (Expeller)	33.0	09.0	0.44	0.89	0.58	1.24
Linseed Meal (Solvent)	34.0	09.0	0.40	0.83	0.60	1.38
Malt Sprouts, Dried	25.0	15.0	0.20	0.70	0.18	0.21
Milo, Grain Sorgham	09.0	02.5	0.04	0.31	0.17	0.34
Molasses, Beet	06.5	00.0	0.10	0.02	0.23	4.77
Molasses, Cane	03.0	00.0	0.82	0.08	0.35	2.38
Oats	12.0	12.0	0.10	0.13	0.17	0.37
Oat Hulls	04.0	30.0	0.09	0.10	0.19	0.40
Peanut Hulls	08.0	63.0	0.25	0.06	0.19	0.82
Peanut Meal (Expeller)	45.0	12.0	0.15	0.55	0.33	1.15
Peanut Meal (Solvent)	47.0	13.0	0.20	0.60	0.04	1.15
Rice Bran	13.0	12.0	0.10	1.50	0.94	1.35
Rice Hulls	04.4	40.0	0.08	0.06	0.10	0.34
Soybeans	38.0	05.0	0.25	0.59	0.21	1.70
Soybean Hulls	12.5	35.0	0.54	0.16	0.70	0.86
Soybean Meal (Expeller)	42.0	06.5	0.20	0.60	0.27	1.97
Soybean Meal (Solvent)	44.0	07.0	0.25	0.60	0.27	1.97
Soybean Meal (Solvent) dehulled	49.0	02.5	0.20	0.65	0.27	1.97
Urea	282.0	00.0	0.00	0.00	0.00	0.00
Wheat, Hard	13.0	02.0	0.05	0.40	0.11	0.32
Wheat, Soft	10.0	02.8	0.05	0.30	0.10	0.41
Wheat Bran	16.0	10.0	0.14	1.17	0.45	1.20
Wheat Middlings	17.0	08.0	0.15	0.91	0.08	0.34
Wheat Shorts	18.4	05.0	0.11	0.76	0.26	0.88
			Minerals			
Bone Meal, Steamed	00.0	00.0	29.0	14.0	00.0	0.0
Defluorinated Phosphate	00.0	00.0	34.0	18.0	00.0	0.0
Dicalcium Phosphate	00.0	00.0	23.0	18.0	00.0	0.0
Limestone	00.0	00.0	38.0	00.0	00.0	0.0
Magnesium Carbonate	00.0	00.0	00.0	00.0	26.0	0.0
Monosodium Phosphate	00.0	00.0	00.0	26.0	00.0	0.0
Double Sulfate of Potassium and Magnesium	00.0	00.0	00.0	00.0	11.0	18.0

Table 12 Feed ingredients—Their composition and energy value.

NE_{milk}		$NE_{maint.}$		NE_{gain}		Metabolizable energy		Roughage factor	Restrictions % of ration	
Mcal		Mcal		Mcal		Mcal		Units/lb	Milk	G&F
(lb)	(kg)	(lb)	(kg)	(lb)	(kg)	(lb)	(kg)		(Maximum)	
0.59	1.31	0.55	1.21	0.29	0.64	0.94	2.07	0.20	5.0	5.0
0.99	2.18	2.07	4.57	1.18	2.62	1.42	3.14	0.00	2.0	2.0
0.86	1.90	0.86	1.89	0.56	1.25	1.25	2.76	0.10	25.0	0.0
0.73	1.61	0.66	1.46	0.42	0.94	1.09	2.42	0.05	15.0	25.0
0.65	1.43	0.59	1.31	0.35	0.76	1.00	2.22	0.20	0.0	10.0
0.80	1.75	0.80	1.77	0.54	1.18	1.17	2.57	0.20	0.0	25.0
0.98	2.15	0.92	2.02	0.60	1.32	1.38	3.05	0.00	0.0	0.0
0.87	1.92	0.80	1.77	0.54	1.18	1.26	2.77	0.20	0.0	0.0
0.37	0.81	0.43	0.95	0.10	0.22	0.66	1.46	0.60	0.0	0.0
0.86	1.90	0.79	1.74	0.53	1.16	1.25	2.76	0.00	0.0	0.0
0.98	2.15	0.70	1.54	0.45	1.00	1.47	3.23	0.00	0.0	0.0
0.77	1.70	0.70	1.54	0.46	1.01	1.15	2.52	0.00	0.0	0.0
0.31	0.69	0.38	0.85	0.04	0.09	0.60	1.32	1.00	20.0	0.0
0.91	2.00	0.83	1.83	0.55	1.22	1.31	2.88	0.10	0.0	0.0
0.97	2.13	0.90	1.99	0.59	1.30	1.38	3.04	0.10	0.0	0.0
0.97	2.13	0.90	1.99	0.59	1.30	1.38	3.04	0.00	0.0	0.0
1.04	2.30	1.00	2.20	0.63	1.39	1.47	3.24	0.00	0.0	0.0
0.85	1.88	0.78	1.73	0.52	1.15	1.25	2.75	0.00	0.0	0.0
0.79	1.74	0.71	1.57	0.47	1.04	1.16	2.57	0.00	0.0	0.0
0.65	1.43	0.59	1.31	0.35	0.76	1.02	2.27	0.20	10.0	10.0
0.86	1.90	0.79	1.74	0.53	1.16	1.25	2.76	0.00	0.0	0.0
0.66	1.46	0.60	1.33	0.40	0.88	0.98	2.17	0.00	10.0	10.0
0.82	1.81	0.77	1.70	0.50	1.11	1.17	2.57	0.00	10.0	10.0
0.77	1.69	0.70	1.54	0.46	1.01	1.14	2.51	0.10	15.0	15.0
0.31	0.69	0.38	0.85	0.04	0.09	0.60	1.32	1.00	0.0	0.0
0.16	0.35	0.19	0.42	0.00	0.00	0.29	0.64	1.00	0.0	0.0
0.89	1.97	0.81	1.80	0.55	1.20	1.29	2.85	0.00	0.0	0.0
0.73	1.62	0.73	1.62	0.48	1.06	1.19	2.63	0.00	0.0	0.0
0.63	1.41	0.59	1.30	0.35	0.77	0.99	2.19	0.15	5.0	5.0
0.16	0.35	0.19	0.42	00.0	0.00	0.29	0.64	1.00	0.0	0.0
1.03	2.27	0.98	2.16	0.62	1.38	1.45	3.19	0.00	5.0	5.0
0.90	1.98	0.74	1.63	0.49	1.08	1.20	2.64	0.00	20.0	20.0
0.90	1.98	0.84	1.85	0.56	1.23	1.30	2.87	0.00	0.0	0.0
0.83	1.84	0.78	1.72	0.52	1.14	1.22	2.68	0.00	0.0	0.0
0.90	1.98	0.84	1.85	0.56	1.23	1.30	2.87	0.00	0.0	0.0
0.00	0.00	0.00	0.00	0.00	0.00	0.00	0.00	0.00	0.5	1.0
0.90	1.99	0.84	1.85	0.56	1.22	1.29	2.84	0.00	0.0	0.0
0.90	1.99	0.84	1.85	0.56	1.22	1.29	2.84	0.00	0.0	0.0
0.68	1.51	0.62	1.36	0.38	0.85	1.03	2.29	0.00	0.0	0.0
0.87	1.93	0.80	1.76	0.53	1.18	1.26	2.79	0.00	0.0	0.0
0.91	2.01	0.85	1.87	0.56	1.24	1.31	2.89	0.00	0.0	0.0
					Minerals					
0.00	0.00	0.00	0.00	00.0	0.00	0.00	0.00	0.00	0.0	0.0
0.00	0.00	0.00	0.00	00.0	0.00	0.00	0.00	0.00	0.0	0.0
0.00	0.00	0.00	0.00	00.0	0.00	0.00	0.00	0.00	0.0	0.0
0.00	0.00	0.00	0.00	00.0	0.00	0.00	0.00	0.00	0.0	0.0
0.00	0.00	0.00	0.00	00.0	0.00	0.00	0.00	0.00	0.0	0.0
0.00	0.00	0.00	0.00	00.0	0.00	0.00	0.00	0.00	0.0	0.0
0.00	0.00	0.00	0.00	00.0	0.00	0.00	0.00	0.00	0.0	0.0

by increasing lactic acid in the silage. Several experiment stations have reported good results from ensiling complete rations (adding the grain to the forage going into the silo) and this idea may have merit in that it permits mechanical feeding of a complete ration.

CONCENTRATES

Because of their wide range in composition, grains and by-products of grain processing have traditionally been used to balance the deficiences of forage in dairy rations. In practice, their chief contributions have been in supplying energy and protein. Prior to the availability of vitamin and mineral supplements, ingredients were also selected for their ability to provide these ingredients. As will be pointed out in future chapters, such items as influence of the ration on rumen fermentation and the relative concentration of the ration and its influence on intake must also be considered in balancing rations. Since there are no miracle feeds, the primary factors to be considered in selecting concentrates for the ration are protein content, energy value, and the limitations imposed by influence on intake and rumen fermentation. The major concentrates used in dairy rations are listed in table 12 along with their composition, energy value and suggested limits on their use.

CONCLUSIONS

The current widespread use of least-cost well balanced rations for dairy cows and the excellent levels of production that have been maintained leaves little doubt that the principles of feed evaluation and selection outlined here are useful and accurate. It goes without saying that future research will further improve ration selection and use. Meantime, use of the principles, values and restrictions outlined in this and other parts of this book will result in economical milk production and animal growth.

References

1. Axellson, Joel. The Relationships in Silage Between pH-Value and Contents of Dry Matter, Crude Fiber, Ash and Butyric Acid. The Annals of the Royal Agricultural College of Sweden. 18:7, 1951.

2. Barnett, A. J. G. Silage Fermentation. Academic Press, Inc. New York. 1954.

3. Conrad, H. R., Hibbs, J. W., Pratt, A. D. and Davis, R. R. Relationships Between Forage Growth Stage, Digestibility, Nutrient Intake and Milk Production in Dairy Cows. Ohio Agricultural Experiment Station Research Bull. 914. 1962.

4. Gordon, C. H., Melin, C. G., Irvin, H. M., and Wiseman, H. G. Variability in the Quality of Untreated High Moisture Silages as Related to the Apparent Success of Preservatives. U.S.D.A. A.R.S.-52-9. 1955.

5. Murdock, J. C. Seepage trom Silos. Agriculture, 61:224. 1954.

6. National Research Council. Joint United States-Canadian Tables of Feed Composition.

7. Shaw, A. O., Golding, W. S., and Ashworth, U.S. Influence of Temperature and Atmosphere on Silage Preservation. J. Dairy Sci. 34:494. 1951.

Chapter 4

The Rumen and Its Use in Nutrition

As indicated in the title of this chapter, our concern will be with using the rumen in the nutrition of the animal rather than the more conventional discussion of the microbiology, chemistry, and anatomy of the rumen found in most books.

Had the dairy cow and her relatives who possess a four compartment stomach been supplied with the enzyme cellulase, all of this would have been unnecessary. The two maior sources of feed energy are starch and cellulose The ruminant, as well as the non-ruminant, are both adapted to using starch since both have an amylase enzyme. In the absence of a cellulase enzyme, the cow was supplied with an amazingly complex stomach and matching physiology to enable her to use cellulose as a major source of energy. Since both starch and cellulose are composed of the same building blocks, and both yield carbon dioxide and water as end products of their breakdown, it can almost literally be said that the difference between the need for a simple and a four-compartment stomach is the presence of a β linkage in cellulose and an α linkage in starch. One can even dream that had the cow been given a cellulase enzyme, our modern dairy cow would have been as sleek as a greyhound and body capacity would have been an undesirable characteristic on the dairy score card.

In fact, however, the cow has no natural means of using cellulose as an energy source and her entire being is adapted to using anaerobic bacteria as her cellulose-digesting tissue. To do this, the four compartment stomach is necessary along with a physiology that is adapted to using volatile tattv acids as a maior source of energv rather than glucose.

To make this system work when the only source of nutrients was the natural plant materials growing in a locality, the ruminant had to be able to do several things for itself. Over the course of a year, the animal would be confronted with a variety of nutritional situations which would require changes in the microorganisms for their exploitation. The rumen is sunnlied with a vast number of bacteria and protozoa which are capable of changing their population ratios to meet the demands of the feed being supplied at the moment. For this reason, all attempts that have been made to date to improve rumen fermentation by feeding additional microorganisms or to alter the microbial population have been without success. Even with the high starch rations fed to today's high producing dairy cows, the microbial population appears quite capable of making any needed changes.

The well-known imbalance of amino acids in plant proteins presented another problem for the ruminant. This too has been taken care of by the simple move of having the microorganisms breakdown the plant proteins and rebuild them into proteins that are well balanced in terms of the needs of the host animal. The true magnitude of this ability of the cow to provide for her protein needs is illustrated in the several instances when all of the nitrogen in the diet has been supplied from urea. In such cases the microorganisms not only had to manufacture all needed amino acids but had to do so with only ammonia as a nitrogen source. It is not surprising therefore that all attempts to show a need for feeding a "balanced" protein in ruminant rations have been unsuccessful.

The third area in which the ruminant had to be self-sustaining was in the manufacture of vitamins. The fat soluble vitamins—A, D, and E—could be stored in times of plenty for times of want, but the water soluble vitamins must be available on a day to day basis. Here again, the ruminant utilizes the microorganisms and their ability to produce water-soluble vitamins in the normal course of fermentation. For this reason, there is no evidence of a requirement for the addition of the so-called water soluble vitamins to the diet of ruminants fed normal rations.

A note of caution should be sounded here. In research a very useful tool in the studies of rumen fermentation has been the artificial rumen. As a means of studying the nutritional requirements of the microorganisms in the rumen, the microorganisms are sometimes removed in near pure culture by high-speed centrification and their growth studied in water solutions with and without varying sources of nutrients. When this is done, it can be demonstrated that the microorganisms require vitamins, amino acids, volatile fatty acids, minerals, and varying sources of energy for their growth. Frequently, such data are used to promote the sale and use of complex "rumen fermentation" supplements containing trace minerals, vitamins, and "balanced" proteins. This is a complete misuse of the research data. The fact that the microorganisms exhibited a need for these nutrients when separated from their normal source of nutrients cannot be extrapolated to indicate a need under normal feeding conditions.

From the above discussion, it can be seen that the ruminant needs a daily supply of only three nutrients: a nitrogen source, an energy source, and a mineral source. A source of vitamins A, D, and E is also required, but it can hardly be listed as a daily requirement since the animal is capable of storing these vitámins and, therefore, of going for short periods without them in the diet.

THE RUMINANT DIGESTIVE SYSTEM

The ruminant stomach is a four-compartment one consisting of the rumen (the large fermentation vat), the reticulum, the omasum, and the abomasum or true stomach. The role of each compartment can be seen best in terms of the basic requirements of a digestive system designed to ferment and utilize the nutrients from large quantities of wet bulky materials (figure 14). Fermentation is a relatively slow process which is aided by frequent mixing and a grinding process to reduce the particle size of the materials and thus increase the surface area available for attack by bacteria. The rumen is ideally suited for this and the act of rumination handles the need for reduction in particle size. Some regulation of the movement of materials from the rumen to the lower tract is required and this is apparently at least one of the functions of the reticulum. The great amount of water needed in the rumen to provide optimum conditions for fermentation would obviously be a great disadvantage to the smaller abomasum, so one of the main functions of the omasum is the removal of water, thus providing the abomasum with a smaller quantity of much concentrated materials for further digestion.

The ruminant stomach therefore provides for the five conditions which are necessary for the continuous culture of microbial populations.

1. A regular substrate supply for the microorganisms is provided by frequent intakes of feed.

2. Inhibition of fermentation by the accumulation of end-products of fermentation is prevented by absorbtion through the rumen wall and movement through the system.

3. Rumen temperature is maintained by body-temperature regulators of the animal.

4. Rumen volume is regulated by almost continual passage of liquid materials to the omasum through the reticulo-omasal orifice.

5. Fluid volume and pH of the rumen is maintained by the secretion of large volumes of saliva.

It must not be concluded that the ruminant digestive system is an entity unto itself. While it is true that the digestive system is a continuous tube

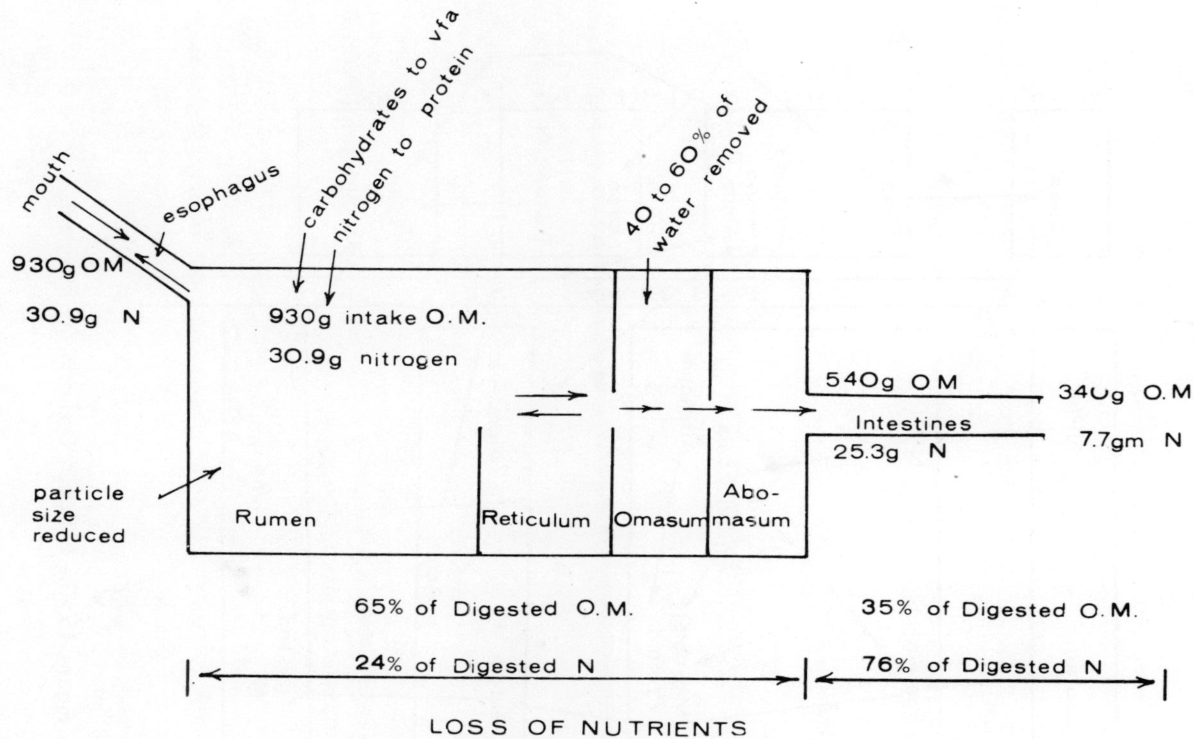

Figure 14 The Movement of Feed thru the Digestive Tract of the Ruminant, the Primary Changes Which Occur and The Relative Loss of Nutrients Above and in the Intestines.

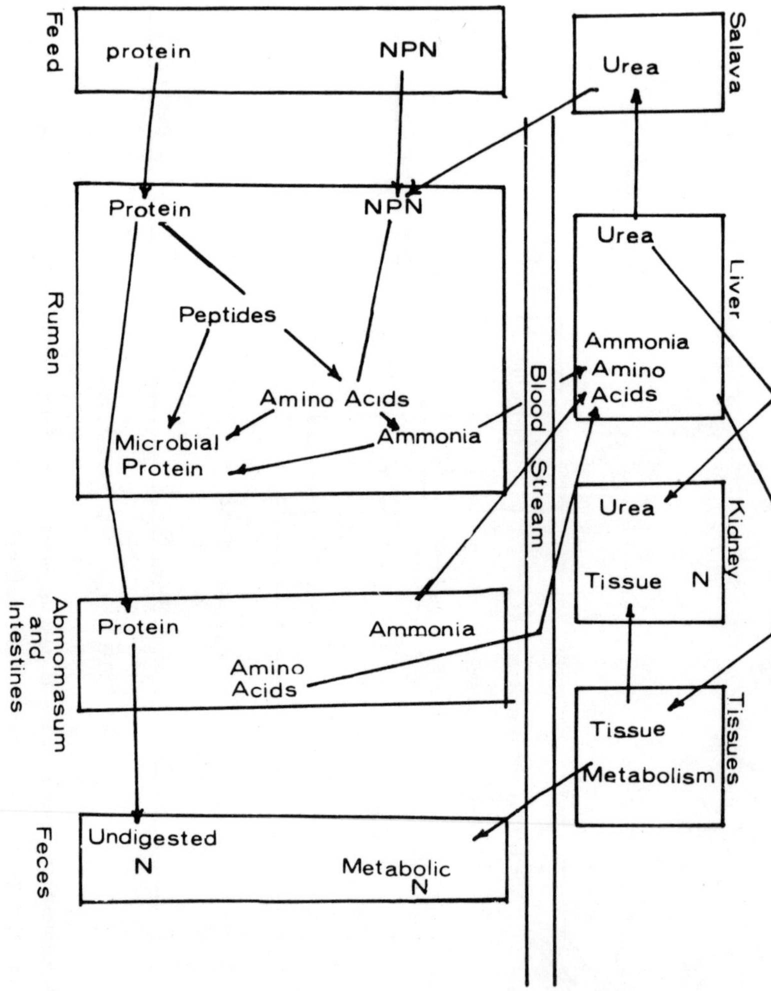

Figure 15 The Nitrogen Cycles in the Ruminant.

from the mouth to the anus without a single opening into the tissues and all passages of materials from the tract to the other organs and tissues is via the blood stream, the function of the rumen is very much related to the function of other organs. This is illustrated in figure 15 in the case of the nitrogen cycle in the ruminant. It can be seen that there is an almost continuous passing of nitrogen in the form of ammonia, urea, and amino acids through the bloodstream between the rumen, kidneys, liver, tissues, and saliva. This flow of nitrogen means that the animal is capable of handling nitrogen in either an excess of needs or to make maximum use of nitrogen when supply barely meets demand. This flexibility is a characteristic of most living organisms and is a natural buffer to the variety of nutritional situations that a cow may be subjected to. As will be seen later, this flexibility in the use of nutrients prevents the statement of exact nutrient requirements for animals and differentiates biological functions from physical functions in which a measured input of energy yields predictable reactions under standard conditions. In short, this flexibility differentiates the cow from an electric motor in terms of using energy to perform its natural functions.

The data in figure 16 further illustrates the complexity of ruminant digestion in terms of the many enzyme systems involved, some of which are produced by the microorganisms and some by the several organs of the animals. Thus the amylase of the rumen may be of bacterial origin while that in the intestines is produced by the body. The same is true of the proteases and the lipases. Despite this complex interaction of the rumen and other organs, the overall suitability of the rumen to handle many complex feeding situations is illustrated by the fact that all attempts to improve general function of the ruminant by adding enzymes to feed have met with failure.

It was mentioned earlier that the ruminant was adapted to utilizing the end products of microbial fermentation as an energy source as opposed to the use of glucose by non-ruminants. This adaptation has resulted in more than an ability—it is almost a necessity. The many problems encountered when all concentrate rations are fed to ruminants are ample evidence of this fact. The end-products of microbial digestion of carbohydrates are methane, carbon dioxide, water, and the fatty acids chiefly lactic, acetic, propionic, and butyric. The process also liberates heat which may be either good or bad, depending upon the atmospheric temperature under which the animal is living. The three end-products that have received primary attention in the nutrition of ruminants are the three volatile fatty acids: acetic, propionic and butyric. The importance of these acids as sources of energy to the ruminant is shown in table 13 which illustrates a typical energy distribution of two rations fed to dairy steers at the Georgia Experiment Station. The rations, as fed, had a calculated content of 28,957 kcal. of digestible energy. Of this energy which was digested and available for use by the animal, between 18,000 and 19,000 kcal. appeared as the three volatile fatty acids. This constituted about 66 per cent of the digestible energy of the rations. On a molar percentage basis, the ratios of the acids was about 65 per cent acetic,

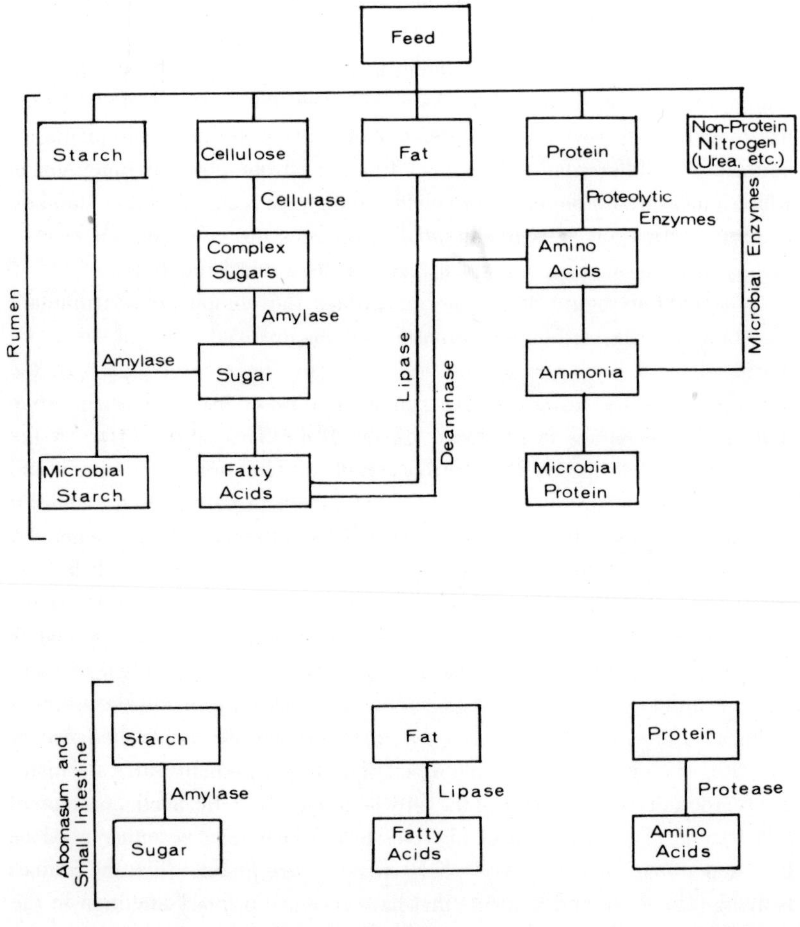

Figure 16 Some of the Known Enzyme Systems that are of Importance in Feed Utilization by Ruminants.

Table 13 The production of volatile fatty acids in rumen fermentation.

Item	Ration 1[a]	Ration 2[b]
Feed intake (kg dry matter)	9.6	9.6
Rumen volume: water, ml	30,000	36,144
dry matter, kg	7.98	9.59
Animal weight, kg	408	408
Mean concentration: (rumen fluid)		
Acetic acid (kcal)	9475	9239
Propionic acid (kcal)	4300	4137
Butyric acid (kcal)	5309	5516
Total energy as acids	19,084	18,892
Energy intake (dig. kcal)	28,957	28,957
Percent dig. kcal as fatty acids	66	65

[a] Corn silage 65%, grain 35%.
[b] Corn silage + cellulase 65%, grain 35%.
McCullough, Georgia Station.

20 per cent propionic, and 15 per cent butyric. This general ratio is normal for rations containing large amounts of highly digestible roughage. On an energy basis, however, about 50 per cent of the kcal. was present as acetic acid, about 20 per cent as propionic, and 30 per cent as butyric acid. It is not surprising, therefore, that the influence of the ration fed on the end-products produced in the rumen is strongly reflected in such animal performance measures as quantity of milk produced, milk fat percentage, efficiency of conversion of feed to milk and relative value of a ration for milk production as opposed to fattening.

The importance of not only the total acid production but the ratio in which the acids are produced in milk production is illustrated in the data in table 14 and figure 17. The data are from experiments in which pure acids are slowly infused into the rumen or added in feed fed at frequent intervals as an additional supply of nutrients. When the acids were added to rations for cows being underfed, all three acids improved the total amount of milk produced. Acetic and butyric acids improved milk fat percentage while propionic acid reduced it. The same acids infused in addition to a normal rate of feeding gave a different picture in terms of milk production. Both acetic and butyric acid improved milk fat percentage while propionic acid reduced it. Total milk production was improved by acetic acid but reduced by propionic and butyric. This data is further illustrated in figure 17 along with an experiment in which the acids were added to a ration being allowed at a level of intake that provided surplus nutrients. The ability of acetic acid to improve both total milk production and milk fat percentage was demonstratable even under conditions of surplus intake. The data leave open a still unanswered question as to the relative importance of the total production of volatile acids vs. the molar percentage ration in which they are produced. This area of research may hold the key to future advances in the efficiency with which feed is converted to meat and milk.

MEANS FOR CONTROLLING RUMEN FERMENTATION

The previous section of this chapter has demonstrated the importance of the rumen and rumen fermentation in animal production. The logical question which follows is "what can be done about it in the feed trough?" At the present time, the feeder has five means available for exercising control over the rumen fermentation process and the end-product which are available for the production of meat and milk. They are:

1. Feed ingredients that are fed.

Table 14 The influence of the infusion of acids on milk production.

Infusion	Yield	Fat	SNF
	(lb/day)	%	%
Infused in addition to normal feed[a]			
Water only	36.0	3.78	8.87
Acetic acid	36.8	4.06	8.83
Propionic acid	33.5	3.57	9.10
Butyric acid	35.1	4.17	8.91
Infused in addition to limited feed[b]			
Water only	20.0	4.62	8.90
Acetic acid	23.2	4.88	8.90
Propionic acid	22.7	4.22	9.18
Butyric acid	22.8	5.79	8.77

[a] Rook, J. A. F., Balch, C. C. and Johnson, V. W. (1965) Further Observations on the Effects of Intraruminal Infusions of Volatile Fatty Acids and of Lactic Acid on the Yield and Composition of the Milk of the Cow. British J. Nutr. 19:93.

[b] Wilson, G. F., Davey, A. W. F., and Dolby, R. M. (1967) Milk Composition as Affected by Intro-ruminal Infusion of Volatile Fatty Acids to Cows on a Restricted Ration. N. Z. J. Agric. Res. 10:215.

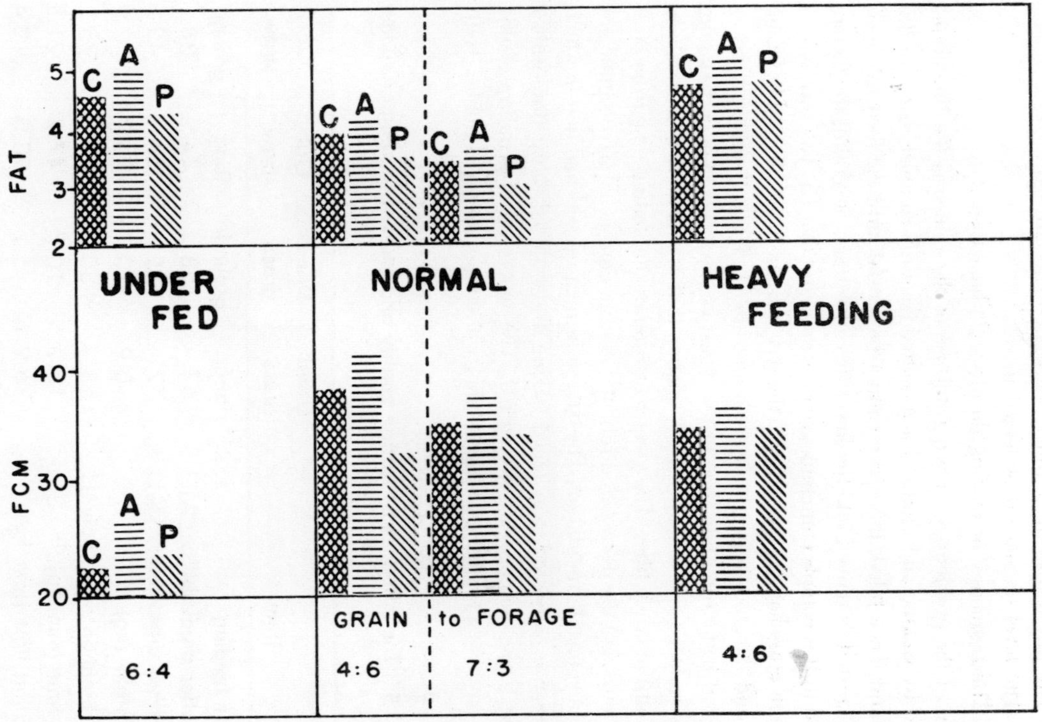

Figure 17 The Interactions Between Plane of Nutrition and the Effects of Either the Infusion or Feeding of Volatile Fatty Acids on Milk Production and Milk Fat Percentage. (C = control, acetic = acetic acid, propionic = propionic acid.)

2. Ratio of ingredients included in the total ration.
3. The physical form in which the ingredients are fed.
4. The level or amount of feed fed.
5. The frequency of feeding during a 24-hour period.

One of the problems in making full use of the rumen is the limitations imposed by volume and intake. Each animal has only so much space in the rumen and, once filled, this space cannot be reused until the feed has been either digested or moved into the lower digestive tract. Several of the factors involved in this complex situation are illustrated in table 15. The two forages used were grass hay and oat straw—the hay being a highly digestible forage (67%) and the straw a poorly digested one (44%). When the grass hay was fed long, it had a retention time in the digestive tract of 53 hours. Grinding the hay reduced the amount of dry matter that was digested, particularly the amount of crude fiber. The great reduction in ruminating time from 36.9 to 5.2 hours and the decrease in reticular contractions could have significant effects on rumen fermentation. The data indicate a rumen condition in which the mass was subjected to reduced mixing and reduced saliva flow with a resulting fermentation at a lowered pH. The net result of grinding of this type of hay was a lowering of digestibility without an improved intake. With

Table 15 The influence of forage quality and form of feeding on rumen function.[a]

Item	Dried grass	Dried grass	Oat straw	Oat straw
Form of feeding	long	ground	long	ground
Dry matter digestibility (%)	67	58	44	38
Crude fiber digestibility (%)	72	45	54	42
Daily intake (kg)	10.9	10.7	5.7	7.2
Mean retention time (hours)	53	50	75	73
Eating time (min/kg)	18.2	10.7	47.2	21.8
Ruminating time (min/kg)	36.9	5.2	93.0	12.4
No. reticular contractions/day	1372	1049	1812	1684
Weight of material transferred from reticulo-rumen per contraction (g)	3.9	6.0	1.9	2.6

[a]Campling, R. C. and Freer, M. Factors Affecting the Voluntary Intake of Food by Cows. 8. Experiments with Ground, Pelleted Roughages. British J. Nutr. 20:229. 1966.

oat straw, similar effects were recorded for digestibility, ruminating time, eating time and fiber digestibility. The grinding process did increase total intake.

The data illustrate the problem of reducing particle size as a means of speeding feeds through the rumen. The ruminant is obviously well equipped to do its own reduction in particle size by rumination. When this is supplanted by grinding, the materials may be speeded through the rumen but a price must be paid in the form of lowered digestibility and an altered fermentation. Since this usually takes the form of producing less acetic and more propionic acid, such a change may be desirable in growing or fattening animals but not for milking cows. The real lesson probably is in the comparison of the two forages. When high quality forage is fed the cow needs no help in obtaining a high level of intake and maintaining a high level of digestibility.

Another factor involved in utilizing the rumen is the influence of level of feeding. The data in table 16 illustrate this in terms of feeding the same

Table 16 The influence of form of feeding and level of intake on digestibility and rate of movement through the digestive tract.[a]

Form in which dried grass was fed	Amount given daily	Apparent digestibility	Mean time spent by food in the digestive tract (hour)
	grams	(%)	
Long	600	80.3	103
	1200	79.1	72
	1500	79.4	68
Medium-ground and cubed	600	76.9	74
	1200	71.5	53
	1500	69.9	42
Finely ground and cubed	600	75.9	53
	1200	68.8	39
	1500	65.4	34

[a] Blaxter, K. L., McCgraham, N., and Wainman, F. W. Some Observations on the Digestibility of Food by Sheep, and on Related Problems. British J. Nutr. 10:69. 1956.

forage in three forms at three levels of intake. At the low level of feeding (600 gm/day) the sheep were obviously underfed and a very high percentage of the feed was digested even when the feed was finely ground and pelletted although the time spent by the feed in the digestive tract was reduced by nearly one-half. Doubling the rate of feeding further reduced the digestibility of all forms and further reduced the time spent in the digestive tract. Going from 1200 gm to a full-feed at 1500 gm continued all trends. The data illustrate very well the overall problem of the feeder that wants to make maximum use of his feeds and yet needs to obtain maximum animal production. The ruminant does a maximum job of digesting feed when it is underfed. This is particularly true of forages with a low percentage of digestible dry matter. On the other hand, to achieve maximum rates of gain or milk production, near maximum levels of intake are required. How to live with this problem in the feedlot or milking barn constitutes a major problem in ruminant nutrition.

In practice, the solution to the problem of an intake of adequate feed to provide high levels of energy, is to feed a combination of forage and grain. When this is done, a whole new set of relationships in the rumen become evident. The data in table 17 show some of the changes. Changing from an all hay ration to a 50:50 ration of hay and grain increased total intake by about one-third while actually reducing the weight of digesta in the reticulo-rumen. The dry matter digestibility of the entire ration was increased reflecting the higher digestibility of the grain portion of the ration. The

Table 17 The influence of ratio of hay to grain on rumen functions.[a]

Item		Daily feed (kg)			
	Hay	10.2	9.9	8.1	7.0
	Grain	0	2.5	5.0	7.5
Total intake of dry matter (kg)		8.8	10.7	11.4	12.6
Water consumption (kg/day)		33	35	36	42
Retention time of hay (hours)		60	61	68	70
Dry matter digestibility (%)		66	67	70	69
Crude fiber digestibility (%)		70	72	60	54
Digesta in reticulo-rumen (kg)		68	68	62	63

[a] Campling, R. C. The Effect of Concentrates on the Rate of Disappearance of Digesta from the Alimentary Tract of Cows Given Hay. J. Dairy Res. (1966) 33:13.

retention time of the hay portion of the ration was increased but fiber digestibility continued to be depressed. The data do illustrate some tendency for the favorable effects of adding grain to forage to be curvelinear in nature and to have an "optimum" point. This point, and the influence of combining forage and grain would be expected to vary with the quality of the forage and the nature of the grain.

In 1965, Montgomery et al., introduced the concept that intake was influenced by a combination of ration density and digestibility and suggested a measure resulting from the multiplication of the two measures. In figure 18 are the results of feeding several feedstuffs calculated in terms of density × TDN. The data illustrate the general influence of several common feeds on intake and show an optimum at a density × TDN of about 35. In practice, one would expect that a feed having a density × TDN value above 35 could be combined with one below 35 with increased intake so long as the density × TDN of the ration did not get above the optimum. The data in figure 19, show the influence of adding two feedstuffs—corn with a density × TDN value of 71 and beet pulp with a value of 28—to corn silage to create rations with several percentages of grain. At a high level of intake with milking cows, the corn + silage rations showed the expected initial increases in intake followed by a decrease at high levels of corn feeding. Adding a low density feedstuff such as beet pulp gave only a small initial increase. With dairy heifers and a lower level of feeding, the influence of corn was less pronounced but the effects of the low density beet pulp were accentuated. The intake data in figure 19 again illustrate the complex inter-actions of forage quality, grain density, level of grain feeding and level of total ration consumption on dry matter intake and, of course, rumen function. The density × dig. (TDN) of several common feedstuffs as determined in the author's laboratory is in table 18.

The importance of using only high quality forage when high levels of intake are desired cannot be over-emphasized. The data in figure 20 again illustrate the point. Using silage harvested from an oat crop at 3 stages of maturity, rations containing several forage to grain ratios were fed to dairy heifers. The initial supremacy in intake of the early-cut, highly-digestible silage was maintained at all levels of grain feeding. Thus, in using the rumen to obtain animal performance, grain cannot substitute for the detrimental effects of low quality forage.

Just as there is an optimum set of conditions for total intake of feed, so there is a set of optimum conditions for maintaining a desired type of fermentation in the rumen. In the previous discussion, the item most fre-

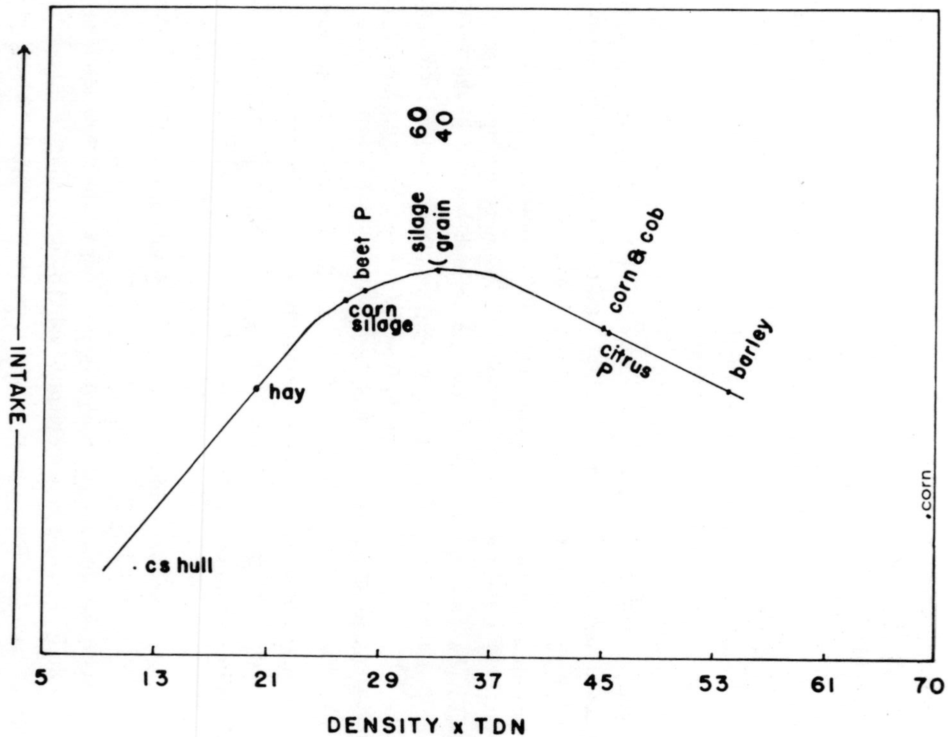

Figure 18 The Influence of Density X TDN on the Intake of Dry Matter.

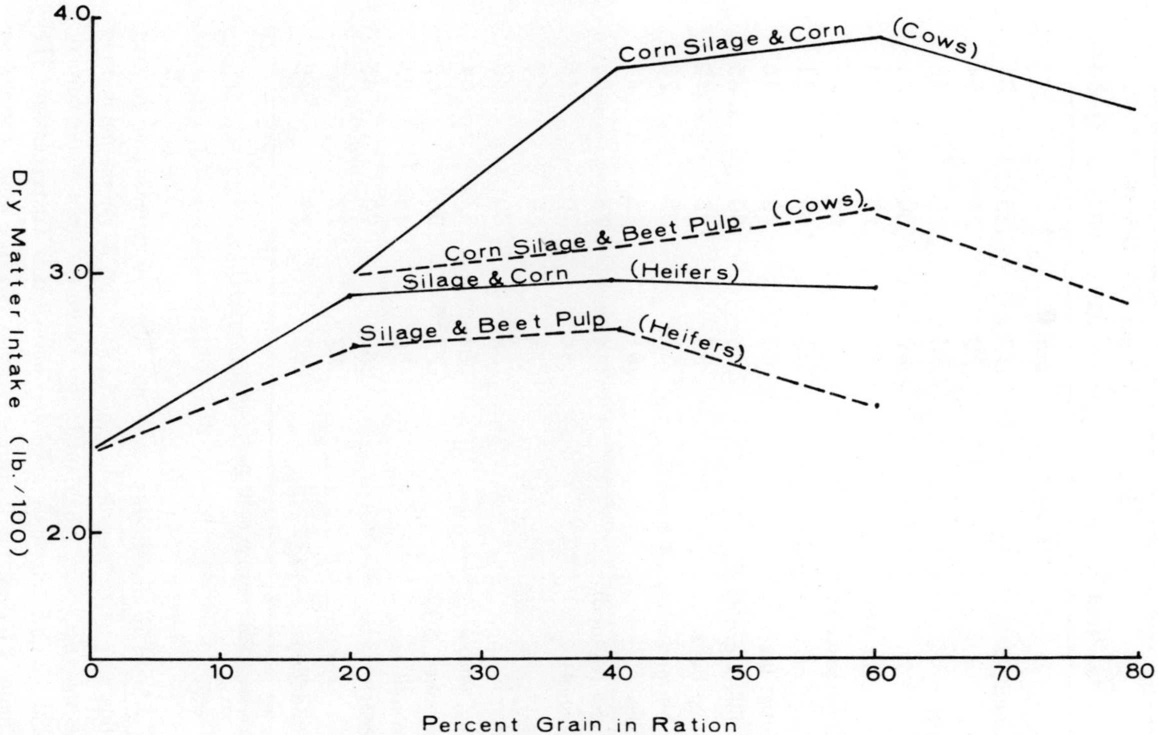

Figure 19 The Influence of Several Ratios of Corn and Silage or Beet Pulp and Silage and Two Types of Animals on Dry Matter Intake.

Table 18 The density × digestibility of some common feedstuffs.[a]

Feedstuff	Weight per ml. (D)	Digestibility	D × Dig.
	gm	%	
Alfalfa meal, dehy.	.57	55	31
Barley	.70	77	54
Beet pulp	.40	70	28
Corn	.89	80	71
Corn and cob meal	.61	73	44
Corn silage (excellent)	.41	65	27
Corn silage 60% + grain 40%	.46	68	31
Corn gluten feed	.63	73	46
Corn distillers grains	.56	80	45
Cottonseed hulls	.33	38	12
Coastal bermuda hay (4 weeks)	.37	55	20
Citrus pulp	.61	75	46
Citrus meal (burned)	.44	75	33
Linseed meal	.64	70	45
Oats	.74	68	50
Sorghum	.54	80	43
Soybeans, whole	.74	87	64
Soybean meal, 50%	.77	80	62
Soybean mill run	.55	45	25
Oat groats	.50	90	45
Wheat	.87	77	67
Wheat middlings	.63	77	48

[a] All samples ground through a 40 mesh screen.

quently reduced in digestibility was crude fiber. In fattening or growing rations this may be of little or no significance, but, in rations for milking cows this may be a very important factor due to the need of the cow for acetic acid for both total milk yield and normal milk fat percentage. The data in figure 21 show the influence of grain to silage ratio on cellulose digestibility in rations containing corn silage and either beet pulp or corn at the Georgie Experiment Station. The data clearly indicate that high levels of grain feeding are detrimental to cellulose digestibility but that the nature of the cellulose present may alter the extent of the influence. The data are used here to introduce the idea of optimum conditions in the rumen to produce a desired type of fermentation.

OPTIMUM FERMENTATIONS

In figure 22, the influence of pH in the rumen on the production of several acids and on animal response are illustrated. It can be seen that

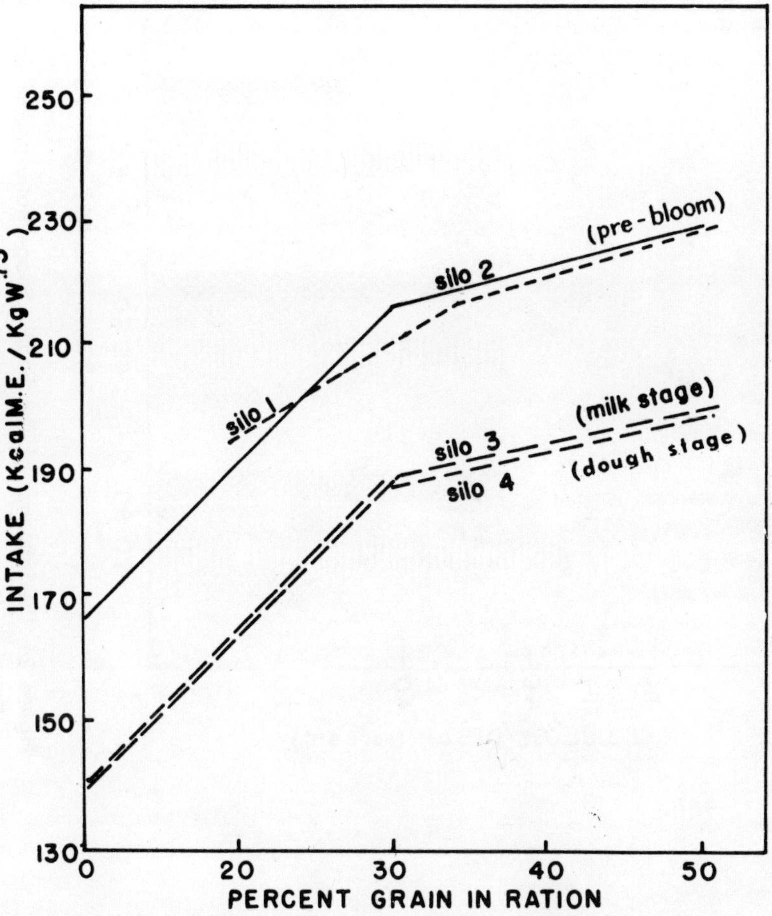

Figure 20 The Influence of Stage of Harvest of Wheat Silage and the Percent of Grain in the Ration on Dry Matter Intake.

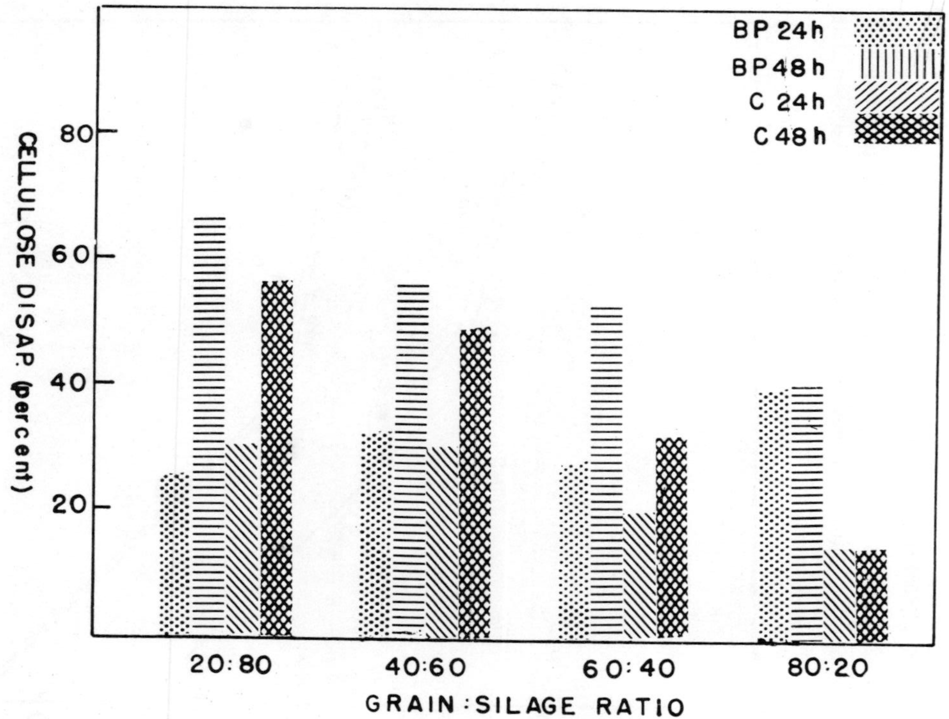

Figure 21 The Influence of The Grain : Silage Ratio and of Corn vs. Beet Pulp on Cellulose Disappearance in Artificial Rumens.

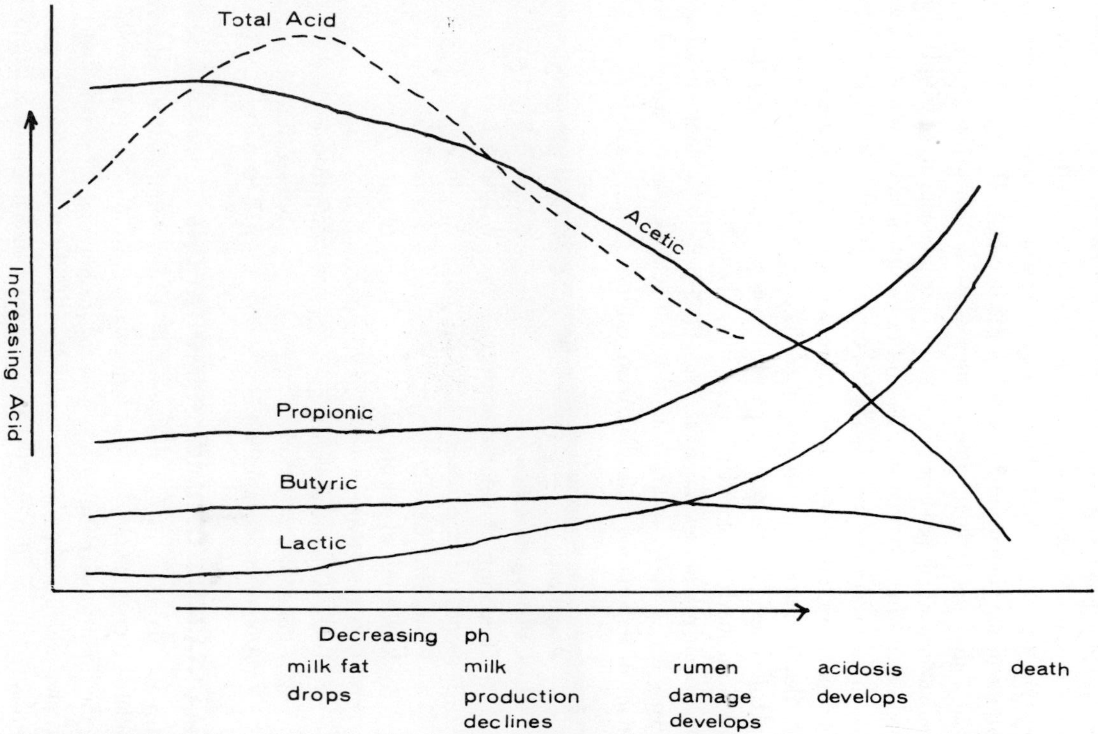

Figure 22 The Overall Relationships between pH of Rumen Fluid and the Development of Abnormalities in Ruminants.

acetic acid production is steadily reduced as the pH decreases with a resulting increase in the production of propionic acid but with little change in the production of butyric acid. Lactic acid is seldom found in measurable quantities in the rumen until a very low pH develops. This may be due to a rapid conversion of lactic acid to other acids at high pH or to a failure of the bacteria which produce lactic acid to predominate at a high pH. The normal chain of events that can be said to be detrimental with decreasing pH are in the order of depressed milk fat, depressed milk production, damage to the rumen wall, development of severe acidosis (severe indigestion) and finally death. Since many factors other than ration composition influence the pH developed in the rumen, no prediction of resulting rumen pH can be made on the basis of ration composition alone.

Optimum pH for several rumen reactions have been determined. The following list is by no means complete, but it does contain many of the reactions of importance to feeding dairy cattle.

Reaction	Optimum pH range
1. Cellulose digestion	6.0 to 6.8
2. Formation of volatile acids	6.2 to 6.6
3. Synthesis of protein	6.3 to 7.4
4. Proteolytic activity	6.5 to 7.0
5. Reduction of nitrate	about 6.5
6. Urease activity	7 to 9
7. Lactate fermentation	5.9 to 6.2
8. Ammonia production	above 6.2
9. Ammonia consumption	5.7 to 6.2

MAINTAINING OPTIMUM FERMENTATIONS

Up to this point in our discussion of rumen fermentation, we have considered the factors which affect fermentation. We now come face-to-face with the problem of outlining a system for maintaining those conditions in the rumen which will insure maximum digestibility of the ration, maximum digestibility of crude fiber, and the maintenance of such functions as ammonia utilization and volatile acid production which are needed for maximum ration utilization. Before outlining the system in theory, it might be useful to look at one in practice.

One of the remarkable examples from research is the demonstration at the National Institute of Animal Husbandry in Denmark of the influence of proper feeding on milk production. The results in table 19 show the before and after milk and fat production of a group of cows fed and managed to

Table 19 Production of cows before and during experimental year.[a]

Cow no.	Milk (lb) Before exp. year	Milk (lb) In exp. year	Butterfat (lb) Before exp. year	Butterfat (lb) In exp. year
187	18091	25754	826	1189
255	16369	25825	770	1130
279	12793	24427	523	1068
242	14482	19833	671	1050
218	14714	20379	671	1038
155	14213	23505	668	1036
294	12824	20900	564	1005
177	14491	21607	651	1001
272	13946	21583	590	929
178	11519	18192	498	854
Average	14345	22200	643	1030

[a] Larson, L. H. and Ezkedal, H. W. Feeding of High-Yielding Dairy Cows. Reprint of 260. Report from the National Institute on Animal Husbandry, Copenhagen, Denmark. 1952.

include the best management and feeding practices available. Their success in increasing milk production from 14,345 to 22,200 and fat from 643 to 1030 pounds cannot be attributed entirely to the system of feeding but, in light of modern knowledge, they certainly followed the proper feeding schedule to get the job done. The daily feeding schedule was as follows:

3:30 a.m. Concentrates followed by water and roots.
4:00 a.m. Milking (a small portion of concentrates fed during milking). After milking, silage and hay.
9:30 a.m. Concentrates followed by water and roots.
10:00 a.m. Milking (a small portion of concentrates). Silage and hay after milking.
3:30 p.m. Concentrates followed by water and roots.
4:00 p.m. Milking (a small portion of concentrates). Silage and hay after milking.
9:30 p.m. Concentrates followed by water and roots.
10:00 p.m. Milking (a small portion of concentrates). Silage and hay after milking.

36% of the feed units fed were from concentrates and 64% from forage.

The workers used a feeding program that derived 36 per cent of the feed units from grain and 65 per cent from forage which means that the forage was of high quality, the density × Dig. was optimum, the feeding schedule maintained an even curve of fermentation, rumen pH was kept high and the production of volatile acids was not only high but the cows obviously had adequate acetic acid for a high level of milk and normal milk fat percentage.

The workers in Denmark were utilizing what is sometimes called a "continuous fermentation" in the rumen in contrast to the initiation of a series of fermentations which accompany many feeding programs. The curve in figure 23 illustrates the influence of frequency of feeding on the stages of fermentation in the rumen. The curve resulting from a single daily feeding represents the several stages in a normal fermentation such as one might use to make alcohol or other fermentation products on a batch basis. Following feeding, there is a lag phase followed by a very rapid acceleration phase. It is in this rapid acceleration phase that the rumen pH may drop to very low levels depending upon the amount of grain in the ration and the ratio of acetic to propionic acid may be altered to produce low milk fat percentage. The acceleration phase is followed by an exponental phase during which the rate of fermentation is relatively constant and it is in this phase that the detrimental effects of prolonged low pH may manifest themselves in rumen damage or acidosis if the ration contains a high starch content and is fed at the high level, thus assuring adequate substrate for a continuing fermentation. Thus at a low level of feeding, the low pH accompanying the acceleration phase would be only temporary and the detrimental effects noted at full feed might not occur. The exponental phase is followed by a retardation phase as the nutrients for fermentation are exhausted and finally a stationary phase when the fermentation is finished.

When the animal is fed twice daily, the acceleration and retardation phases are reduced and with four times per day feed, they are largely eliminated.

The net effect of altering this curve of fermentation on several measures of feed utilization is demonstrated in the data in table 20 which summarizes work by Moir and Somers with sheep. Feeding the same ration and the same total quantity of feed in four feedings increased the digestibility of the ration, increased the amount of protein stored, reduced the range of pH in the rumen, reduced the amount of free ammonia in the rumen and reduced the maximum number of rumen bacteria. The influence of frequent feeding on the protozoa is of particular interest. This group of rumen microorganisms is of particular

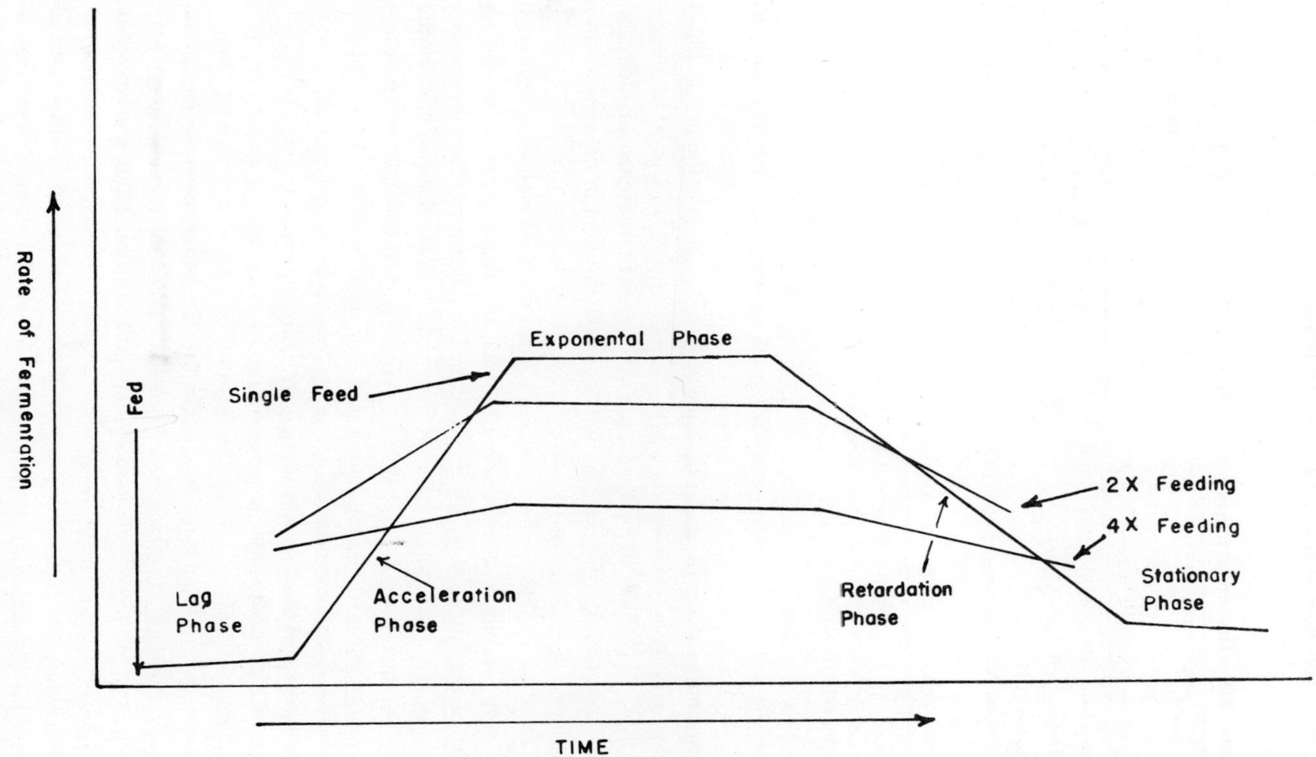

Figure 23 The Influence of Frequency of Feeding on the Rate and Shape of The Fermentation Curve in the Rumen.

Table 20 The influence of frequency of feeding on ration utilization.[a]

Item	Frequency of feeding		
	1×	2×	4×
Dry matter digestibility (%)	63.9	66.1	67.1
Nitrogen balance (g/day)	2.29	2.78	3.18
Range in pH	5.8–7.4	6.6–7.2	6.7–7.4
Maximum NH_4 nitrogen (mg/100 ml)	16.8	15.3	11.6
Maximum rumen bacteria ($\times 10^9$/ml)	41	30	35
Mean number of protozoa ($\times 10^9$/ml)	1.15	2.26	3.14

[a] Moir, R. J. and Somers, M. Ruminal Flora Studies VIII. The Influence of Rate and Method of Feeding a Ration upon the Digestibility, upon Rumen Function, and upon the Ruminal Population.

interest when rations containing high levels of starch are fed. The protozoa have the ability to absorb starch granules and thus take them out of circulation in the rumen fluid. By doing this, they not only aid in maintaining a high rumen pH, they also keep the ratio of acetic to propionic acid at a more favorable level. Since the protozoa are very sensitive to rumen pH, frequent feeding is about the only way we have to maintain them in the rumen when high levels of grain must be fed.

The data in table 21 illustrate another plus for frequent feeding. No small part of the problem in maintaining a high pH is the presence of enough buffer in the rumen in the form of sodium and potassium. Since the primary source of these minerals is the saliva, the data are of interest in that normal levels of minerals were maintained in the rumen even with rations containing 80 per cent corn when they were fed in four equal feeds at 6-hour intervals.

In practice, dairy cows are supposedly fed at something approaching twice daily feeding. Actually, with most dairy herds this is not done. With high levels of milk production, large quantities of grain are fed at milking time and forage in between. When the quantity of grain fed at milking time is large enough to reduce the cow's appetite for forage, cows frequently do not eat forage for some hours after being fed grain. In effect, what we have is four separate fermentations during the 24-hour period: two primarily of grain and two primarily of forage. The frequently observed improvement in fat test that results from feeding part of the grain on top of silage instead of in the barn attests to the failure of conventional feeding techniques to promote "optimum" fermentation.

The data in table 22 illustrate the effects of feeding the same total ration in four equal feeds at 6-hour intervals. Feeding the grain portion separate from the forage portion significantly increased the amount of propionic acid formed and extended the range of pH in the rumen. When the grain was fed as one of the four feeds, the production of propionic acid was further

Table 21 The concentration of minerals in rumen fluid of steers fed silage and four ratios of grain.

Grain : silage ratio	Steer no.	Time	parts/million in rumen fluid Na	K	Mg	Ca	P
20:80	1	8 a.m.	1063	1100	94	25	9
		10 a.m.	1212	900	60	21	16
		12 N	850	800	49	19	13
	2	8 a.m.	1125	700	41	22	13
		10 a.m.	1250	750	84	19	15
		12 N	1212	750	26	17	19
40:60	1	8 a.m.	1162	650	75	19	18
		10 a.m.	1375	650	16	17	9
		12 N	1025	750	26	21	15
	2	8 a.m.	1250	800	52	25	15
		10 a.m.	1250	750	68	22	16
		12 N	1212	650	26	18	22
60:40	1	8 a.m.	1162	800	18	21	18
		10 a.m.	1062	800	20	19	19
		12 N	1525	700	23	19	23
	2	8 a.m.	1025	900	53	24	16
		10 a.m.	1212	900	33	21	24
		12 N	1525	750	26	22	21
80:20	1	8 a.m.	1162	1200	66	14	25
		10 a.m.	1212	950	74	13	24
		12 N	1162	950	47	9	24
	2	8 a.m.	1062	900	40	14	21
		10 a.m.	1125	1050	68	18	23
		12 N	1250	900	32	21	18

McCullough, Georgia Station.

Table 22 The influence of feeding schedule on rumen fermentation.

Item	Time	4 equal feedings	¼ grain 7am–7pm ½ silage 1pm–1am	All grain 7am ⅓ silage 1pm–7pm–1am
Acetic acid (M%)	8 am	66	65	63
	10 am	66	70	62
	12 N	67	65	61
	2 pm	65	63	59
Mean		66[a]	66[a]	61[b]
Propionic acid (M%)	8 am	18	19	20
	10 am	17	17	20
	12 N	17	20	22
	2 pm	19	22	24
Mean		18[a]	20[b]	23[c]
Butyric acid (M%)	8 am	11	12	12
	10 am	12	10	13
	12 N	12	11	13
	2 pm	12	11	13
Mean		12[b]	11[a]	13[b]
Total acid (mg/100 ml)	8 am	538	572	491
	10 am	605	634	603
	12 N	515	530	748
	2 pm	600	626	658
Mean		564	590	625
Rumen pH	8 am	6.5	6.6	6.7
	10 am	6.6	6.5	6.4
	12 N	6.8	6.6	6.3
	2 pm	6.4	6.2	6.2

[a] Means marked with different superscript characters are significantly different.
McCullough, Georgia Station.

increased, the production of acetic acid decreased and the curve of fermentation took on a shape similar to "batch" fermentation.

When carried to the extreme, this "slugging" of animals with grain can lead to the well known condition called "acidosis," acute indigestion or founder. Although seldom seen in dairy herds, this condition is frequently encountered in fattening steers on very high levels of grain and in dairy herds being fed so-called "complete rations." The date in table 23 illustrate the normal progress of this condition under experimental conditions. In practice, it is most frequently observed when animals which are hungry are suddenly exposed to a full feed of a ration containing over 60 per cent grain. In the feedlot, a steer may go along for days doing quite well on such a ration and then suddenly develop the problem and die. When this happens, it usually means that the steers missed a meal or two for some reason and then suddenly ate a large quantity of feed. One case seen in a dairy herd involved cows that had been without feed for 12 hours because the feed truck was late. When the truck came and the hungry cows ate heavily of the grain ration, several cows developed the classic symptoms of acidosis in a few hours.

Using the rumen to its maximum in animal nutrition will require all of the energies that nutritionists and feeders can put to the task. It must be obvious that no set of rules can be formulated that will cover all situations. When feeding programs are developed, the following questions should be kept in mind if the rumen is to function properly.

1. What type of animal performance is required? Rations for milk production present problems which are different from rations for fattening.

Table 23 The development of typical acidosis.[a]

Day No.	Time	Rumen pH	Comment	Rumen acids Percent of total		
				Acetic	Propionic	Lactic
1	9:30 am	6.8	Prefeeding	69	18	1.5
1	9:45 am	—	3 lb. wheat	—	—	—
1	Noon	6.0		58	20	7.7
1	4:00 pm	6.0		57	18	7.5
1	8:30 pm	5.9		53	18	3.7
2	9:00 am	6.3		47	21	5.3
2	9:30 am	—	3 lb. wheat	—	—	—
2	1:00 pm	5.7	Feces soft	35	9	47.1
2	4:00 pm	5.0	Rumen contents very fluid	—	—	—
3	10:00 am	4.7	Ruminal stasis	9	4	85.0
3	12:15 pm	4.3	Profuse diarrhea	—	—	96.0
4	8:00 am		Dead			

[a] Ryan, R. K. Concentration of Glucose and Low-Molecular-Weight Acids in the Rumen of Sheep Following the Addition of Large Amounts of Wheat to the Rumen. American J. Vet. Res. 25:646. 1964.

2. Do the ingredients in the ration require processing to maximummize rumen fermentation or intake? Routine processing should not be done unless it will serve a purpose.

3. Can the ration utilization be improved by feeding? When rations are being fed for maintenance or low levels of animal performance, there is little to be gained from frequent feeding.

4. Should the grain and forage be fed separately or as a mixed total ration? When high levels of grain are being fed, feeding a mixed total ration will assure maximum use of the forage.

CONCLUSIONS

The basic mechanisms of the ruminant digestive tract have been outlined and many of the factors involved in its use discussed. The point cannot be over-stressed that it is a complex system and involves many complex interactions. Difficulties in using the rumen as a part of the process for making maximum utilization of feed for animal production come from feeding systems which accentuate the normal processes in the rumen. The rumen can best be used when a curve of fermentation which closely resembles "continuous fermentation" is maintained. The use of this curve in developing feeding programs will be the subject of later chapters of this book.

References

1. Blaxter, K. L., McGraham, N., and Wainman, F. W. Some Observations on the Digestibility of Food by Sheep and on Related Problems. British J. Nutr., 10:69. 1956.

2. Campling, R. C. The Effect of Concentrates on the Rate of Disappearance of Digesta From the Alimentary Tract of Cows Given Hay. J. Dairy Res. 33:13. 1966.

3. Campling, R. C. and Freer, M. Factors Affecting the Voluntary Intake of Food by Cows. 8. Experiments with Ground, Pelleted Roughages. British J. Nutr. 20:229. 1966.

4. Larson, L. H. and Ezkedal, H. W. Feeding of High-Yielding Dairy Cows. Reprint of 260. Report from the National Institute on Animal Husbandry. Copenhagen, Denmark. 1952.

5. Moir, R. J. and Somers, M. Ruminal Flora Studies. VIII. The Influence of Rate and Method of Feeding a Ration Upon the Digestibility, Upon Ruminal Function, and Upon the Ruminal Population. Aust. J. Agric. Res. 8:253. 1957.

6. Montgomery, M. J. and Baumgardt, B. R. Regulation of Food Intake in Ruminants. 2. Rations Varying in Energy Concentration and Physical Form. J. Dairy Sci. 48:1623. 1965.

7. Rook, J. A. F., Balch, C. C. and Johnson, V. W. Further Observations on the Effects of Intraruminal Infusions of Volatile Fatty Acids and of Lactic Acid on the Yield and Composition of the Milk of the Cow. British J. Nutr. 19:93. 1965.

8. Ryan, R. K. Concentration of Glucose and Low-Molecular-Weight Acids in the Rumen of Sheep Following the Addition of Large Amounts of Wheat to the Rumen. American J. Vet. Res. 25:646, 1964.

9. Wilson, G. F., Davey, A. W. F., and Dolby, R. M. Milk Composition as Affected by Intra-Ruminal Infusion of Volatile Fatty Acids to Cows on a Restricted Ration. N. Z. J. Agric. Res. 10:215. 1967.

Chapter 5

Combining Nutrition and Physiology for Growth and Fattening

One of the more expensive operations on dairy farms is the raising of replacements for the dairy herd. Most dairymen prefer to raise their own replacements to prevent the introduction of diseases into their herds and to make full use of their breeding programs. The value of raising replacements rather than buying them has been documented many times and the widespread adoption of the practice by dairymen attests to the soundness of the program. Few areas of dairying have seen so great a change in practice during the last 25 years as the general philosophy of raising replacements.

In the era of 1930 to 1950, there developed a school of thought that the newborn dairy calf was a very delicate single-stomached animal that must be handled and fed much like a new-born baby. On several dairy farms in the United States, a special barn with incubators was built and the new-born calf was hustled to this "nursery" to be subjected to a form of treatment totally outside of today's possibilities. It was generally understood that this newborn calf presented very difficult problems in nutrition which could only be solved with very complex rations which contained a variety of proteins including some of animal origin, all known and unknown vitamins and, of course, a complete mineral mix. The writings of some of the people who had been successful calf raisers in the era make interesting reading in our era. Much of the success of these people was probably due to the successful avoidance of calf-hood diseases which are now controlled with antibiotics. The calf is a very versatile animal and readily adapted to both the "hospital" atmosphere and the "pablum" formulas used in the period.

In agriculture, we have gone through many periods that are best termed fads. In part, these fads are the result of special economic conditions which exist from time to time. It was one of these fads that resulted in much of the change in practice in calf feeding and management. Beginning in the late 1930's and extending into the 1950's much of the dairy world went through an era that is frequently called the "grassland era." The era received much of its encouragement from World War II with its influence on labor

Table 24 Incidence of rumination by Jersey and Holstein calves up to six weeks of age

Age at observation periods	Jerseys			Holsteins		
	Total	Ruminating		Total	Ruminating	
days	no.	no.	%	no.	no.	%
6–14	11	7	63.6	14	11	78.6
20–28	11	10	90.9	14	14	100.0
34–42	11	11	100.0	14	14	100.0

supply and equipment for cultivation. During this era, pasture and permanent grasses were the thing. In many areas of the world including much of the diary country on the North American continent, any farmer using clean cultivated crops felt it necessary to apologize for his antiquated practices. Beginning in New Zealand and finally covering the globe, researchers began to try their hand at raising young dairy calves on pasture. This "revolutionary" development quickly divided the dairy world into the usual "for" and "against" camps. As pointed out earlier, the dairy calf is an adaptable animal and it made the change from inside to outside with little more than a typical "stiff-legged" jump. The ability of the young calf to adapt to a program which either prolongs or shortens his period as a non-ruminant obviously depends upon the opportunity that is provided for rumen development.

The literature on calf raising which was written in the 1950's is largely a collection of papers proving that raising dairy calves is really a very simple task requiring only the use of simple but high-quality feeds and good management to ward off disease. Typical of the research results from this era are those shown in tables 24 and 25. The data in table 24 is part of a Tennessee

Table 25 Response of calves fed an all dry feed at an early age (to 63 days of age)

Item	28 day group	8 day group	15 day group
Milk replacer consumed (lbs)	22.7	21.8	22.4
Starter consumed (lbs)	115.8	112.3	116.2
Hay consumed (lbs)	25.1	22.0	21.7
TDN consumed (lbs)	113.5	108.6	111.7
Average Daily Gain (lbs)	0.86	0.70	0.91

study which demonstrated the short period of time required for the calf to begin rumination when it was supplied with good hay and a simple grain mixture in addition to its milk replacer. Many experiments showed that a calf given access to high quality forage from birth will digest feed as thoroughly as an adult animal by six weeks of age. The data in table 25 illustrate a typical finding on the ability of the calf to derive most of its nutrition from dry feed after two or three weeks of milk replacer feeding. This data from a North Carolina experiment compared three treatments: (1) milk replacer fed as a liquid through 28 days of age, (2) milk replacer fed as a gruel for 4 days and as a dry pellet from 4 to 28 days, and (3) milk replacer fed as a liquid for 5 days, gruel for 6 days, and as a dry pellet from 4 to 28 days of age. The excellent growth and feed consumption of the calves on ration three shows the ability of the calf to grow normally under very simple feeding programs so long as it has access to simple but high quality rations.

What is now known to be true for the calf, is equally true for the growing heifer. The data in table 26 is typical of research and farmer results and illustrates the two principles of feeding dairy heifers: the need for a ration which supplies high quality roughage and a simple grain mix to balance the deficiencies of the forage. The data are from a New York trial covering three winter seasons during which an early cut hay was compared with a late cut hay and a simple grain ration consisting of 2 parts corn and cob meal + 1 part soybean meal was compared with a complex grain ration consisting of 6.6 parts soybean meal, 1.5 parts cane molasses, 1.5 parts dehydrated alfalfa meal, 0.2 parts bone meal, 0.2 parts trace mineral salt and a vitamin A and D supplement. The data clearly indicate that the requirements for growth

Table 26 Average daily gains of heifers fed two qualities of hay and a simple vs. a complex grain mixture

Hay	Early-cut			Late-cut		
Concentrates	None	Simple	Complex	None	Simple	Complex
			pounds			
Year 1	1.31	1.64	1.56	0.71	1.22	1.32
Year 2	0.94	1.46	1.61	0.90	1.26	1.38
Year 3	1.35	1.64	1.61	0.59	1.13	1.28
3-year av.	1.20	1.58	1.59	0.73	1.20	1.33

and development of the dairy heifer are adequately met by a ration which supplies the needed nutrients in a simple combination and indicated no special need for special sources of energy, protein, minerals or vitamins.

NORMAL GROWTH AND DEVELOPMENT

The question of "what is normal?" is far easier to ask than to answer. The answer is of vast importance to the dairy industry since it will determine the flexibility that a farmer has in growing "normal" heifers at a minimum feed cost. There are at least five alternative methods for feeding growing animals, each of which may permit special economies in replacement costs depending upon the economic conditions which exist at the moment.

1. Steady growth from birth to maturity. This occurs when the animal is given a uniform and ample feed supply.

2. Storage of fat in a period of abundant feed supply to carry the animal through a period of scant feed supply without a serious interruption of growth.

3. Some prolongation of the growth period. This would be achieved by continual feeding of a low plane of nutrition and depending upon time to permit final development of the animal.

4. A compensatory increase in the rate of growth in a period of liberal feeding following a period of low nutrition and slight gain.

5. A reduction in growth rate. This assumes that an animal that is maintained on a low plane of nutrition for a prolonged period of time will adjust to a more economical growth rate than when more liberally fed.

There is little doubt that the first four systems of growth have been and are now in use in raising dairy replacements throughout the world. It is also true that no one of the possibilities can be said to be superior to the other. The problem is to understand the principles involved in each along with the limitations imposed so that the more economical of the systems can be employed when conditions dictate.

THE ECONOMY OF ENERGY CONVERSION

Ritzman and Colovos stated the factors to be considered as follows: "The economy of food energy conversion by farm livestock is contingent on three conditions: (1) the time at which this process is performed; (2) the character of the product stored, and (3) the relative amount of overhead involved." The data in table 27 illustrate that importance of the time in the life of a growing animal that growth is permitted. The young calf has both a high basal metabolism rate and great physiological urge to grow. The data indicate

Table 27 The basal metabolism and body weight increase of Holstein heifers at monthly intervals from birth to 30 months of age.[a]

Age	Weight of animal	Basal metabolism per kg per 24 hrs	Weight increase per day actual	Weight increase per day per 500 kg
	kg	cal	kg	kg
Birth	43			
8 days	52	64	1.125	11.84
Months:				
1	62	49	0.63	6.03
2	94	39	0.82	5.26
3	119	39	1.04	4.89
4	146	36	0.96	3.63
5–7	148–188	34–30	0.76	2.25
8–10	210–281	27–23	0.90	1.83
18–20	363–403	19–19	0.60	0.78
23–30	441–495	18–18	0.22	0.25

[a] Ritzman, E. G. and Colvos, N. F. New Hampshire Agr. Expt. Sta. Tech. Bull. 80. 1943.

the calf in its first few days is capable of increasing in weight at a rate equivalent to a mature animal weighing 1100 pounds gaining 25 pounds a day. This urge to grow declines gradually so that by the time the animal is 18 to 20 months of age its maximum rate of growth is about 1.5 pounds per day. It follows that great economies in the use of feed can be realized by making maximum use of the urge to grow of the young calf providing this rapid growth rate does not have detrimental effects on later performance.

The influence of the character of the product stored refers primarily to whether the animal is storing protein as muscle or energy as fat and also to the nature of the fatty tissue being stored. This will be considered in detail later, but Blaxter has calculated that the caloric value of gains (kcal/kg) when an animal is storing 50 per cent of its energy as fat is 2780, while the value when 90 per cent of the stored energy is stored as fat is 6357 or an increasing cost per unit of weight gain of 2.3 times as much energy. Since this storage of a high proportion of the energy as fat also requires the use of a ration containing a high ratio of grain to forage, it is obvious that fattening is far more expensive than growth.

The third factor—the relative amount of overhead—is frequently overlooked in feeding programs. This cost of maintaining the animals being grown or fattened can easily cancel all the economies realized in the early growth

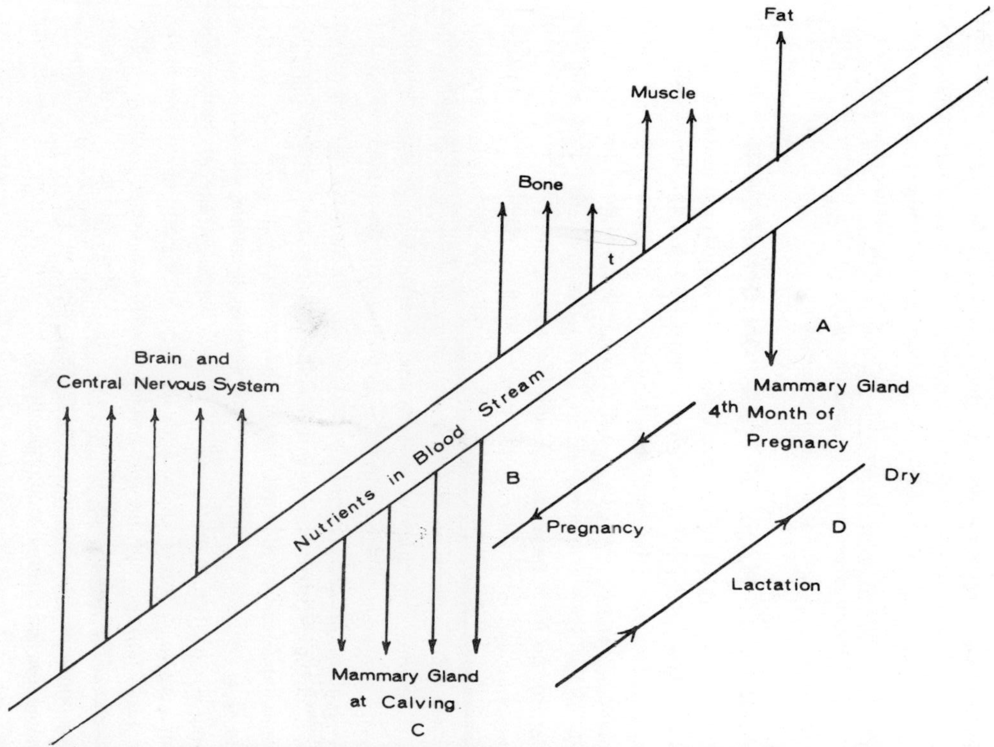

Figure 24 Priority of Partition of Nutrients According to Metabolic Rate. (Dr. John Hammond, British Agricultural Bulletin, Vol. IV, No. 16)

stages of the animal. When an animal reaches 500 pounds of body weight, daily maintenance requirements begin to about equal the requirements for daily growth. When this happens, half or more of the total daily ration goes for maintenance and represents a total loss of feed. Since this cost of maintenance increases with body size, it is obviously very important that the dairy heifer be changed from a consumer of profit to a producer of profit by becoming a milking cow as early as possible.

INFLUENCE OF PLANE OF NUTRITION ON GROWTH

Many authors have reviewed the influence of inanition and malnutrition on growth. Among the living species, vertebrates seem the most resistant to inanition. However, the resistance may vary with species, age, sex, and the individual. Among the factors influencing the resistance, mention may be made of the reserve material of the body. The nervous system and the skeleton are the most resistant, the adipose tissues the least.

Hammond and his students (McMeekan, Wallace, Palsson, Verge) have by the aid of the dissection method greatly extended the knowledge of growth and its relation to nutrition. Hammond's theory of differential growth deals with the general principles of growth and development of farm animals. Relatively, the order of development is as follows: (1) brain and the central nervous system, (2) bone, (3) muscle, and (4) fat. The partition of the supplied nutrients on the different tissues is determined by the order of development. The general priority of nutrient use is illustrated in figure 24.

The data in figure 25 (Palsson and Verges), illustrate the influence of high and low planes of nutrition the first 41 weeks of age in lambs. The level of nutrition resulted in the lambs fed the high plane of feeding being heavier in all portions but the differences were in keeping with the priorities listed above. The ratio of skeleton growth was 1.88 to 1 in favor of the high plane of nutrition. In contrast to this narrow range of relative development of the skeleton, the ratio of subcutaneous fat was 9.94 to 1 and inter-muscular fat 4.65 to 1. It must be explained that high and low plane of nutrition refers only to intake of energy. There is no data to indicate that protein, minerals or vitamins can be fed below their required levels without permanent damage to the animal.

The data in figure 26 (Palsson and Verges) illustrate the influence of energy level on carcass development when animals are permitted to attain equal body weights. The vital organs such as the heart, lungs, liver and abomasum show little difference in weight each having developed in relation to the requirements of the animal. The rumen which develops with age and change in ration was larger in the animals on the low plane of nutrition since the animals were much older at the same weight. The amount of caul and mesenteric fat was also greater in the low plane of nutrition animals again showing the change due to aging. Thus, plane of nutrition may well affect the conformation of an animal but at equal body weights it will have little effect on the development of the vital organs of the body of the skeleton. In practice, therefore, the nutrition of the animal can be altered to meet existing economic conditions as long as the ration is adequate in protein, minerals, and vitamins as well as sufficient energy to permit the development of the vital organs, the skeleton and muscles. The primary point at which

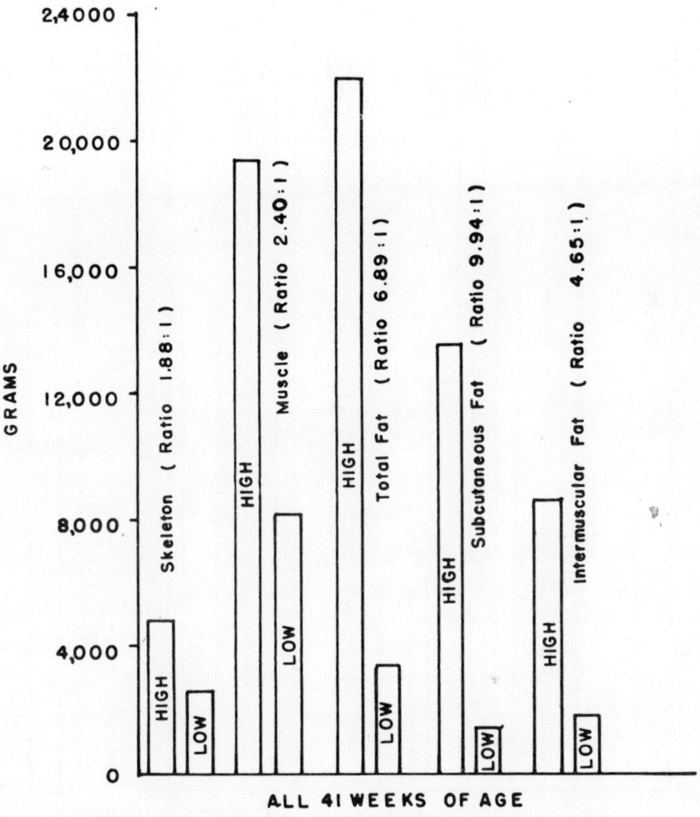

Figure 25 The Influence of a High and a Low Plane of Nutrition on the Relative Increase of the Skeleton, Muscle, and Fat in Lambs From Birth to 41 Weeks of Age.

nutrition can be altered is in the level of energy supplied and the rate at which growth takes place.

There have been many practical experiments throughout the world designed to evaluate the short and long-time effects of plane of nutrition on replacement heifers. Typical of these projects is the work by Reid and

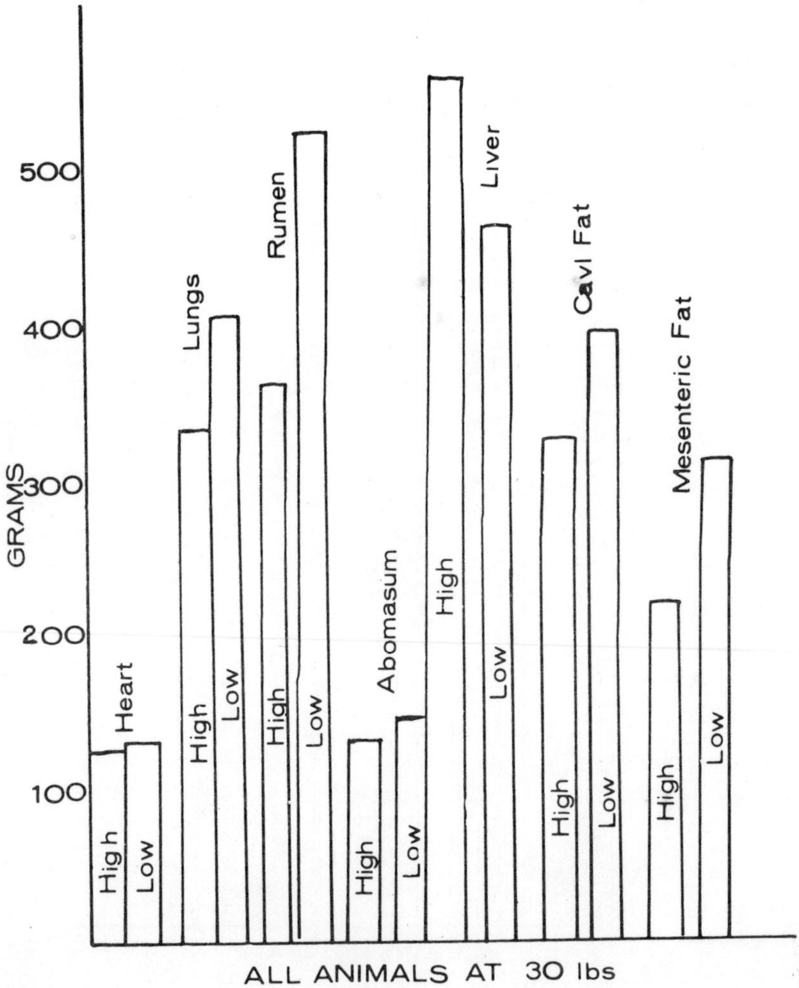

Figure 26 The Influence of Two Planes of Nutrition on The Development of Lambs from Birth to 30 Pounds Body Weight.

co-workers at Cornell. The experiment was designed to measure the influence of three levels of nutrients provided during early life upon the lifetime performance of dairy cattle. The three levels of TDN (as percentages of the upper level of the Morrison feeding standards) fed from birth to the time of first calving were designated as follows: low, 65; medium, 100; and high, 140. The animals were all fed at a high level of nutrition following calving. The data in table 28 summarize the influence of the levels of feeding from birth to two years of age. The results are in keeping with the principles established in the classic experiments of the Cambridge group. The animals varied in body weight from 744 lbs. for the low feeding level to 1253 lbs. for the high level. The several measurements of skeletal growth were in keeping with general body size but were much less affected than body weight. The data in table 29 show the effects of the three levels of feeding on several

Table 28 Body measurements of heifers fed different levels of nutrients from birth to first calving.[a]

Feeding level	Age		
	6 mos.	1 yr.	2 yrs.
Body weight (lbs)			
Low	267	447	744
Medium	339	629	1023
High	449	766	1253
Heart girth (in)			
Low	43.9	53.5	63.5
Medium	48.2	59.6	71.3
High	52.6	63.6	76.2
Height at withers (in)			
Low	36.8	42.7	48.4
Medium	39.3	45.7	50.8
High	42.2	47.6	52.0
Body length (in)			
Low	35.3	41.3	49.5
Medium	38.2	48.1	53.8
High	42.4	51.2	57.9

[a] Proceedings, Cornell Nutrition Conference, 1953.

Table 29 Age and weight of cows at calving time and certain data dealing with reproduction.[a]

Feeding level prior to time of 1st parturition	Age at calving	Services/ conception	Body weight of cow at parturition		Calves			
					Males		Females	
			Before	After	No.	wt.	No.	wt.
	(mos)		(lbs)	(lbs)		(lbs)		(lbs)
First Parturition								
Low	32.1 (15)[b]	1.33 (15)	945 (10)	840 (15)	7	87.1	8	77.1
Medium	29.2 (20)	1.64 (22)	1188 (17)	1067 (20)	7	89.1	13	79.8
High	29.0 (19)	1.89 (19)	1372 (15)	1220 (19)	7	94.3	12	89.5
Second Parturition								
Low	44.0 (10)	1.50 (12)	1415 (10)	1266 (10)	2	101.5	8	92.9
Medium	41.6 (12)	1.53 (15)	1430 (9)	1272 (12)	7	102.4	5	95.8
High	42.9 (11)	1.57 (14)	1571 (11)	1417 (11)	3	105.3	8	98.5
Third Parturition								
Low	55.6 (4)	1.50 (4)	1537 (3)	1347 (3)	2	115.5	1	104.0
Medium	54.8 (7)	1.22 (9)	1526 (7)	1341 (7)	2	106.0	6	87.7[c]
High	55.3 (3)	1.60 (5)	1677 (3)	1530 (3)	2	105.0	1	

[a] Proceedings, Cornell Nutrition Conference, 1953.

[b] Figures in parentheses represent number of observations.

[c] One set of twins included.

measures of reproductive performance. The low level of nutrition tended to delay the onset of puberty and thus the date of first calving. When the animals were changed to a normal level of nutrition after calving, the animals rapidly caught up in body weight, age and subsequent calvings, and other measures of reproduction. The influence of level of nutrition from birth to first calving on subsequent milk production is illustrated in the data in table 30. The fact that the cows grown on the low level of nutrition tended to produce more milk than the other groups in both the first and second lactation is taken to indicate that the low plane of nutrition during rearing did not influence the milk production of the cows. There is a limited amount of information in the literature which indicates that high levels of nutrition which result in the deposition of fatty tissues in the developing udders of heifers can in fact greatly reduce the amount of milk produced after calving. The old idea among successful dairymen that they preferred to have heifers develop udders during the first lactation and not before calving probably has much merit. In practice, heifers should not be permitted to deposit fat in the udder no matter how favorable the current economic climate for fast growth.

In farm practice it is highly unlikely that heifers will be fed on a constant level of nutrition through the growing period. For this reason a New Zealand study comparing several possible combinations of feeding is of special interest.

Table 30 Milk and fat yields of cows receiving three levels of nutrients from birth to first calving.[a]

Feeding-level prior to 1st calving	Number of cows	Days in milk	Milk yield (lbs)	Fat yield (lbs)	% Fat
First lactation					
High	15	314	8890	337	3.79
Medium	16	312	8763	327	3.74
Low	13	359	9496	357	3.76
Second lactation					
High	9	309	10376	369	3.56
Medium	9	288	9801	373	3.80
Low	7	305	11485	422	3.67

[a] Proceedings, Cornell Nutriton Conference, 1953.

The four planes studied were:

Before first calving	After first calving
High	High
Low	High
High	Low
Low	Low

At the first lactation, the well-fed animals (high-high) had the highest yields. The yields for all lactations over a 10-year period were:

	No. of lactations	Lbs. of milk fat
High–high	242	347
Low–high	235	345
High–low	233	308
Low–low	234	305

The high maintenance requirement of heifers during the last year of their growth and the need to have replacements in the milking string as soon as possible to reduce the time from birth to milk production would seem to indicate that on most dairy farms replacements should be grown as rapidly as possible without fattening and without the necessity for substituting large quantities of grain for forage in the ration. To do this, it is suggested that Holstein heifers be fed to average 1.50 pounds per day from birth to calving which would result in heifers weighing 1100 to 1200 pounds at 24 months of age and that heifers of the small breeds be fed similar rations which will result in similar rates of growth. In terms of the feeding practices to be recommended to accomplish this level of growth, normal growth will be defined as an average rate of growth during the first two years of life of 1.25 to 1.50 lbs. per day depending upon the breed of the heifer.

RATIONS FOR NORMAL GROWTH

At this point, we come face to face with the whole problem of nutrient allowances or feeding standards. Over the last 100 years, many investigators have yielded to the urge to reduce all of the nutritional and physiological factors and the range of individuality of animals into neat little tables of allowances for protein, energy, vitamins and minerals for growing and fattening animals. In each case, the tables of feeding standards were carefully labeled as being guides to feeding or suggested average allowances and not absolute values. Unfortunately the labels are almost immediately forgotten and reams of words were written in defense of a favorite standard and great sums of money are spent in an effort to prove that the standard is either good or bad. In the field, the standards get used as absolute values for designing feeding programs. The standards even get programmed into computers and long treatises are written outlining minute differences in feeding programs that will result in amazing economies in farm operations. The simple fact that must be faced is this: the complex problems of feeding animals cannot be summarized in a single value. At best nutrient requirements can only be stated as ranges that cover differences due to feed, animals, and the interactions between the two. It is not surprising therefore that the several standards that are available for consideration sometimes differ by as much as 100 percent. The several protein allowances are shown diagramatically in figure 27. It can be seen that for a 1000 pound dairy animal, the standards allow from a low of 0.45 lbs. of digestible protein to a high of 1.3 pounds. Equally wide differences in suggested allowances can be found for minerals, vitamins and energy. One of the factors involved in the differences is the matter of expected level of ration intake; another the question of "normal" growth; still another, the type of ration fed in terms of protein : energy ratio.

In the author's opinion the more realistic approach to the problem of stating allowances for ruminants is that taken by the "Working Party on Nutrient Requirements of Ruminants" of the British Agricultural Research Council. What follows in the way of specific recommendations has been adapted from the above summaries of estimated requirements.

Energy Allowances There are several factors which affect the amount of energy required for growth, pregnancy, and milk production. Among these factors are the concentration of energy in the diet, the level of feeding employed, the type of animal response desired and the method of expressing energy concentration. The data in figure 4 showed the relationship between the efficiency of utilization of metabolizable energy for maintenance, body gain, and milk production and the concentration of metabolizable energy per unit of ration. It can be seen that maintenance requirements are largely independent of energy concentration, milk production is related in a curvilinear fashion with an optimum use at a concentration of 2.3 to 2.6 Mcal/kg and growth is lineally related with a doubling in efficiency of utilization between a low and a high level of concentration. The data in table 31 show the relationships between the two factors and the net availability of metabolizable energy for maintenance and fattening with 7 ratios of hay and corn in the diet. In preparing the suggested nutrient allowances, the British workers utilized the following factors:

1. The metabolizable energy values ascribed to individual feeds as those

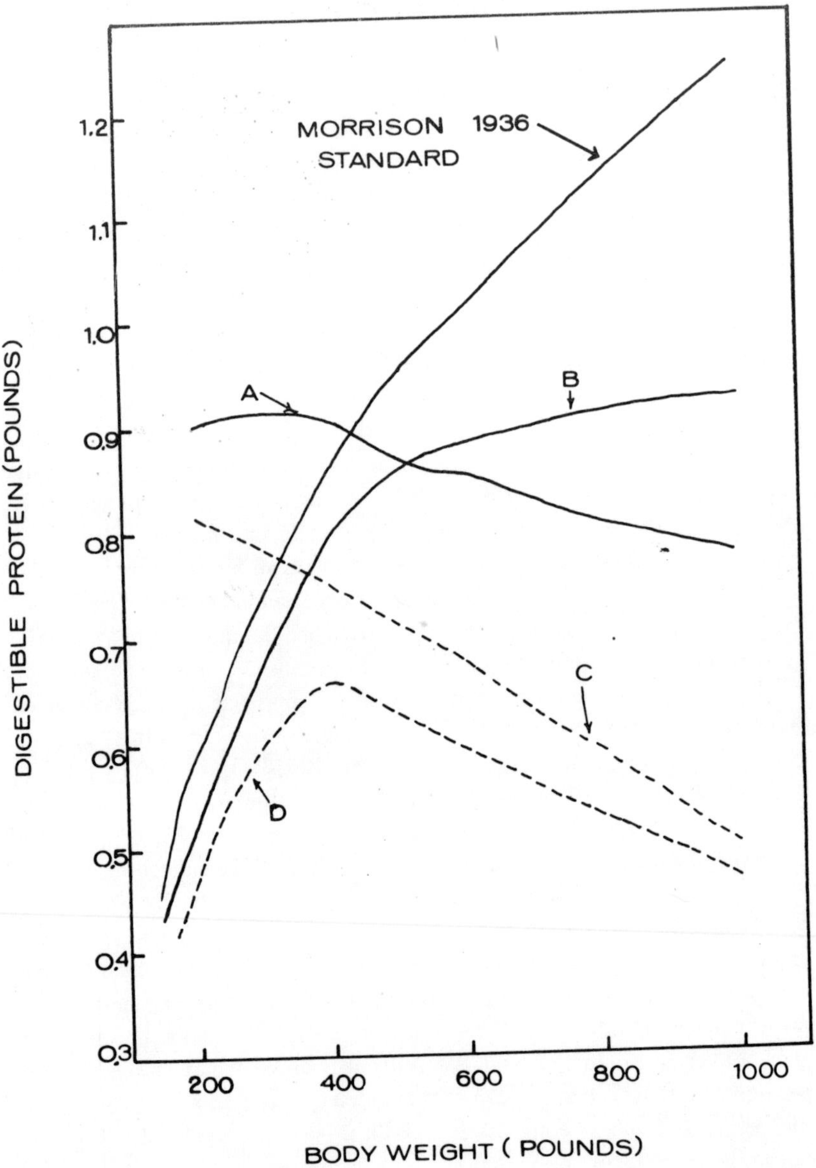

Figure 27 Comparison with the Morrison feeding standard for Protein and Four Other Standards. (Lofgreen, et al, J. Animal Science, Vol. 10, No. 1)

Table 31 Mean net availability of metabolizable energy (kcal./100 kcal.) for maintenance and for fattening in steers and sheep in relation to the proportion of maize in the ration.

Diet. Percentage of		Net availability of metabolizable energy			
		For maintenance		For fattening	
Hay	Maize	Sheep	Cattle	Sheep	Cattle
100	0	70.4	70.0	27.5	31.4
80	20	73.3	74.4	35.4	32.8
60	40	69.7	74.9	44.0	42.1
40	60	74.7	79.3	47.1	47.6
20	80	77.6	81.1	53.3	54.2
5	95	—	77.3	—	63.5
0	100	77.5	—	58.8	—
Standard error of means		±3.8	±1.7	±3.5	±2.8

[a]Blaxter, K. L. and F. W. Wainman. J. Agric. Sci. 63:113.

determined at the maintenance level of feeding, and when diets are calculated, the metabolizable energy to be expected at the particular level of feeding adopted is calculated from the realtionshi ,etween feeding level and the metabolizable energy of the diet as a whole.

2. The estimates include a classification of diets by concentration of energy per unit thus recognizing that the type of diet has a major effect in determining its nutritive value and that individual feeds can be assessed only as components of diets. It is also recognized that the productive value of a feed cannot be given an absolute expression, a fact which is pertinent to problems concerned with the substitution of one feed for another.

Protein Allowances The protein allowances include a consideration of the following factors known to influence the use of protein and the amount of protein needed by the animals.

1. The losses of nitrogen as endogenous nitrogen in urine and feces.

2. The relative retention of nitrogen for live-weight gain by animals of varying weights and at varying degrees of fattening.

3. The additional nitrogen required for pregnancy.

4. The general biological values of nitrogen in ruminant rations.

5. A minimum of 9 to 10 per cent protein required in the ration for normal rumen function regardless of the lower amount which may be indicated by animal requirements.

Mineral Allowances The requirements for the major mineral elements were determined by a factoral method. This involved estimating the net requirement for the element, that is the sum of the amounts inevitably lost from the body (the endogenous loss) and the amounts stored in the body in growth or pregnancy. The net requirement divided by the availability of the element in the diet is an estimate of the amount that must be supplied in the diet.

Intake Allowance This is a major departure from many of the older or current allowances. The allowances take into consideration the several characteristics of rations which influence voluntary intake. Most important the allowances include the influence of ration concentration on intake. Thus, it is recognized that voluntary intake is limited by both high and low concentrations of energy per unit of feed and an optimum level of intake with rations containing between 40 and 60 per cent grain.

Nutrient Allowances for Normal Growth The daily allowances, ration composition required and calculated level of intake for growing dairy heifers gaining at an average rate of 1.50 pounds (.7 kg) per day from birth to 1200 pounds are presented in table 32. The important item in the table is the suggested composition of the rations at each stage of growth. It is suggested that rations be formulated to meet the composition indicated and that they be fed and the resulting rate of performance measured. Should the ration provide for a faster rate of growth than is desired, the rate of growth can be altered by either reducing the rate of feeding or by decreasing the concentration of energy in the ration. Conversely, should the ration fail to provide for the desired rate of growth, it will be necessary to increase the concentration of energy in the ration to provide for both increased intake and increased concentration of energy in the feed consumed. The proposed feeding program would appear to be more in keeping with the more exacting measures of performance needed in modern dairy operations. Since the rations are designed to cover rather wide ranges in body weight, it should be necessary to re-evaluate the rations and feeding programs at 30-day intervals.

NUTRIENT ALLOWANCES FOR FATTENING

Fattening as a production practice is of limited interest in the dairy industry. In recent years there has been renewed interest in some areas in the production of veal and the practice of carrying dairy steers into a fattening stage for beef is increasing. For this reason, some consideration must be given to this area of nutrition. From the general principles outlined earlier and the low priority of fattening on the scale of nutrient uses, it is obvious

Table 32 Suggested nutrient allowances and ration composition for dairy heifers.

Nutrient	Body weight (kg) Body weight (lb)	44 88	50 110	100 220	200 441	300 661	400 882	500 1102
SUGGESTED NUTRIENT ALLOWANCES								
Protein (gm. digestible/day)		150	190	197	210	230	250	270
Energy (kcal/day)		DEPENDS ON RATION CONCENTRATION						
Calcium (gm/day)		12	14	21	24	27	32	33
Phosphorus (gm/day)		8	9	10	13	18	27	31
Vitamin A (IU/day)	When pregnant	—	—	—	—	13,500	18,000	22,500
	Growth	960	1200	2400	4800	7200	9600	12,000
Vitamin D (IU/day)	When pregnant	—	—	—	—	3000	4000	5000
	Growth	160	200	250	500	750	1000	1250
COMPOSITION OF THE RATION								
Protein (%)		—	15	10	10	10	10	10
Energy (Mcal/kg)		3.0[a]	2.6[b]	2.4	2.2	2.2	2.2	2.2
Calcium (%)		—	.75	.75	.36	.36	.36	.36
Phosphorus (%)		—	.40	.40	.30	.30	.30	.30
Vitamin A (IU/lb.)	When pregnant	—	—	—	—	900	900	900
	Growth	600	600	400	400	450	450	450
Vitamin D (IU/lb.)	When pregnant	—	—	—	—	180	180	180
	Growth	100	100	50	50	45	45	45
Calculated Rate of Feeding (lb/day)		—	—	6.0	13	16	20	24

[a] Milk or milk replacer only

[b] Grain ration portion of total ration.

that fattening occurs only when the total intake of energy is in excess of all requirements for maintenance, growth and any other non-fattening function.

Because of the great urge of the young calf to grow which is accompanied by an equally high rate of basal metabolism, fattening the young calf presents special problems in ration formulation. In practice it has been found necessary to utilize liquid diets which maintain the calf as a non-ruminant since the young ruminant fed ruminant-type rations cannot consume sufficient energy to produce fattening. By maintaining the calf as a non-ruminant, and by increasing the caloric content of the liquid ration by including up to 30 per cent animal fat, sufficient energy intake may be achieved to produce considerable fat deposition in addition to normal growth and development. Some idea of the problems involved can be obtained from Blaxter's figures which indicate that calves 2 to 3 months of age storing 60 per cent of their retained energy as fat requires 4307 kcal/kg of gain. Including this amount of energy in 8 liters (2 gallons) of milk replacer in addition to the energy required for maintenance and growth obviously presents a considerable challange to ration formulators. Maintaining a young calf as a non-ruminant presents special problems in preventing digestive disturbances and the stress involved lessens resistance to infections. It is not surprising that the successful production of fat veal requires superior management and special techniques in manufacture of the rations.

The fattening of dairy-type steers presents few problems that are different from the fattening of other steers. The dairy steer has considerable advantages over many beef-type steers because of their superior appetite which permits the use of rations with lower per unit energy concentrations. The primary relationships involved in fattening are those which accompany the change in composition of the fat stored with advancing age and stage of fattening. The daily gains which correspond to given energy retentions in animals of increasing live weight are shown in table 33. A daily gain of 1 kg or 2.25 lb. requires a daily energy retention of about 2400 kcal in young calves weighing 100 kg (220 lb.), 3500 kcal when the animal weighs 300 kg, and 4500 kcal when the steer weighs 500 kg. Thus, doubling the weight of the animal doubles the energy required per unit of weight gain in addition to nearly doubling the requirements for maintenance. These increased requirements are obviously outside of the increase in energy intake which can be expected to accrue from increased intake. The solution must be found in increasing the concentration of energy per unit of ration. In practice, the magnitude of the changes in energy concentration that are required must be understood and ration formulations changed to meet a nearly four-fold

Table 33 Daily gains corresponding to given energy retentions.

Energy retention kcal/day	Live weight of animal (kg)				
	100	200	300	400	500
			Daily gain (kg)		
500	0.26	0.22	0.19	0.18	0.15
1000	0.49	0.42	0.36	0.31	0.28
1500	0.69	0.59	0.52	0.45	0.41
2000	0.88	0.75	0.66	0.58	0.53
2500	1.04	0.90	0.79	0.69	0.63
3000	1.19	1.03	0.91	0.80	0.73
3500	1.33	1.15	1.02	0.90	0.83
4000	1.45	1.27	1.12	1.00	0.92
4500	1.56	1.37	1.21	1.08	1.00
5000	1.67	1.47	1.30	1.16	1.08
5500	1.76	1.56	1.38	1.24	1.15
6000	1.85	1.64	1.46	1.31	1.22
6500	—	1.72	1.53	1.38	1.28
7000	—	—	1.59	1.44	1.34
7500	—	—	1.66	1.50	1.40
8000	—	—	—	1.60	1.47

increase in energy requirement. To illustrate the magnitude of the change required, it might be useful to examine the rations required by a 400 kg (882 lb.) animal for different levels of performance.

Desired Performance	**Ration Concentration**	**Ration Composition**
Maintenance	1.8 Mcal/kg	Daily intake of 6.9 kg (14.2 lb) good forage dry matter.
Maintenance + .25 kg (0.55 lb) gain	1.8 Mcal/kg	8.4 kg or 18.5 lb dry matter of good forage.
Maintenance + .75 kg (1.65 lb) gain	2.2 Mcal/kg	Daily intake of 8.4 kg (19.4 lb) of a mixture of 75% forage and 25% 25% grain.
Maintenance + 1.25 kg (2.75 lb) gain	2.8 Mcal/kg	8.0 kg (17.6 lb) of a mixture of 60% grain and 40% forage.

The requirements indicate that changing the same animal from a maintenance ration to one resulting in about maximum fattening requires a change from an all forage ration fed at something less than full intake to a ration consisting of at least 60 per cent grain fed at maximum intake.

In the absence of reliable tables of nutrient allowances for the entire range of body weights and levels of expected gain, the data in table 34 can be used to calculate requirements in terms of new energy. In calculating allowances for fattening as in allowances for growth and development, the author suggests that the procedure be to calculate rations that can be expected to be consumed in sufficient quantities to permit the desired rate of fattening and the adjustment of energy concentration within the ration to control performance rather than attempting to meet some specific nutrient intake.

Table 34 Net energy requirements[a]

	NE required				NE required		
	For maintenance	For production per pound of gain			For maintenance	For production per pound of gain	
Body weight		steers	heifers	Body weight		steers	heifers
lb	megcal. per day			lb	megcal. per day		
400	3.85	1.34	1.49	**800**	6.47	2.25	2.51
425	4.02	1.40	1.56	**825**	6.62	2.31	2.57
450	4.20	1.47	1.63	**850**	6.77	2.36	2.63
475	4.37	1.53	1.70	**875**	6.91	2.41	2.69
500	4.55	1.59	1.77	**900**	7.06	2.46	2.74
525	4.72	1.65	1.83	**925**	7.21	2.52	2.80
550	4.88	1.70	1.90	**950**	7.36	2.57	2.86
575	5.05	1.76	1.96	**975**	7.50	2.62	2.91
600	5.21	1.82	2.02	**1000**	7.65	2.67	2.97
625	5.38	1.88	2.09	**1025**	7.79	2.72	3.03
650	5.53	1.93	2.15	**1050**	7.93	2.77	3.08
675	5.69	1.99	2.21	**1075**	8.07	2.82	3.13
700	5.85	2.04	2.27	**1100**	8.21	2.87	3.19
725	6.01	2.10	2.33	**1125**	8.35	2.91	3.24
750	6.16	2.15	2.39	**1050**	8.49	2.96	3.30
775	6.32	2.20	2.45	**1175**	8.63	3.01	3.35

[a] Lofgreen, G. P., and Garrett, W. N., Net Energy Requirements, Feed Values for Growing and Finishing Cattle—1967 Revision. Feedstuffs, July 22, 1967.

CONCLUSIONS

The nutritional problems involved in rearing dairy replacements are relatively simple. Since the cost of rearing replacements is rather great, there is considerable need for using all available nutritional "know-how" to keep costs at a minimum. Numerous experiments have indicated that variations in nutrition and growth rate prior to first calving are without influence on milk production so long as the feeding programs do not injure the growth and development of vital organs and the skeleton. This permitted variation applies only to energy: there is no data to support an abnormal feeding of protein, minerals or vitamins.

It is suggested that feeding programs be developed through ration formulation rather than feeding standards. The individual animal will do a better job of determining its relative needs than can be done by using some average value from a table of feed requirements.

References

1. Agricultural Research Council. The Nutrient Requirements of Farm Livestock. No. 2—Ruminants. London, 1965.

2. Blaxter, K. L. and Wainman, F. W. The Utilization of the Energy of Different Rations by Sheep and Cattle for Maintenance and for Fattening. J. Agric. Sci. 63:113. 1964.

3. Hammond, John. Physiological Limits to Intensive Production in Animals. British Agr. Bull. 4:221.

4. Lofgreen, G. P., Loosli, J. K. and Maynard, L. A. Comparative Study of Conventional Protein Allowances and Theoretical Requirements of Growing Holstein Heifers. J. Animal Sci. 10:171. 1951.

5. Lofgreen, G. P. and Garrett, W. N. Net Energy Requirements, Feed Values for Growing and Finishing Cattle—1967 Revision. Feedstuffs, July 22, 1967.

6. Merrill, W. G., Tuck, K. L., Trimberger, G. W. and Loosli, J. K. Value of Simple and Complex Supplements with Early and Late Cut Hay for Yearling Dairy Heifers. Cornell Agr. Exp. Station Bull. 994. 1964.

7. Mochire, R. D. and Musley, W. R. Changing Dairy Calves onto All Dry Feed at an Early Age. J. Animal Sci. 1957.

8. Palsson, H. and Verges, J. B. Effects of the Plane of Nutrition on Growth and the Development of Carcass Quality in Lambs. Part l. The Effects of High and Low Planes of Nutrition at Different Ages. J. Agri. Sci. 42:2. 1951.

9. Reid, J. T. Effects of Several Levels of Nutrition Upon Growth, Reproduction, and Lactation in Cattle. Proceedings, Cornell Nutrition Conf. 1953.

10. Ritzman, E. G. and Colovos, N. F. Physiological Requirements and Utilization of Protein and Energy by Growing Dairy Cattle. New Hampshire Agr. Exp. Sta. Tech. Bull. 80. 1943.

11. Steensberg, V. Feeding Requirements of Young Cattle. British J. Nutrition, 1:139. 1948.

12. Swanson, E. W. and Harris, J. D., Jr. The Development of Rumination in the Young Calf. J. Dairy Sci. 1958.

Chapter 6

Optimum Rations for Milk Production

There are no miracle feed ingredients for milk production. If so, why all the fuss about preparing rations for milking cows? This statement and this question have appeared so often that many dairymen wonder if the statement is true and some research men seem to believe that if they continue to run feeding trial after feeding trial comparing one feed ingredient with another a true miracle will come forth. On today's dairy farms with large herds and even larger investments, milk production must be at a high level both on a per cow and per worker basis. Under these conditions, individual attention to cows must be held at a minimum but at the same time production must be pushed higher and higher. A miracle ingredient that would automatically improve milk production would be most welcome. To date the fact stands—there are no miracle ingredients. This does not mean that ingredients cannot be found and combined in such a way that the ration most nearly meets the nutritional requirements of the herd or that a single combination of ingredients will work best in the same herd year after year or in two neighboring herds the same year. Our purpose here is to outline the basic principles of feeding dairy cows for high levels of milk production as they are now understood, and to provide a framework within which ingredients may be combined to do the best job in a given herd under carefully defined feeding conditions.

WHAT INFLUENCES EFFICIENT PRODUCTION?

Efficient milk production is obviously only partially a function of proper nutrition. The entire range of herd management is important but outside the scope of this book. Within the items that may operate in a given herd under a set level of management skills, the obvious are genetic limitations of the cows, level of feeding, and balance of ration nutrients.

The data in tables 35 and 36 illustrate the relative importance of the three items in an experiment in Sweden. The object of the research was to compare the influence on milk production of several ratios of forage and grain when fed at equal levels of nutrition (equal intake of Scandinavian Feed Units), when fed at three levels of nutrition and when fed to cows with three levels of ability to produce milk. The data in table 35 represent the results when the ration was composed of hay and grain and the hay was of poor quality thus representing a feeding regime that was borderline for maximum milk production. The data in table 36 illustrate a similar comparison but with the forage coming equally from hay and silage thus providing, in this case, a better overall feeding regime.

The importance of the production capacity of the cows is clearly evident in both trials. The spread in average milk production between the cows with a high and a low capacity to produce milk is greater than the differences due to any other factor studied. When the data from the two trials are compared, it is apparent that the cows with a high production capacity were limited somewhat on the poorer feeding regime. When the superior rations

Table 35 Average feed intake and production of cows fed hay and grain during a 189 day feeding trial.

Comparison	Feed units fed	Ration crude fiber	Average daily milk	Pounds of milk per feed unit	Change in body weight
		%	lbs.	lbs.	lbs.
Hay : Grain ratio					
70:30	10.22[a]	27.6	34.5	1.54	−35
55:45	10.38	25.1	35.4	1.55	+27
40:60	10.89	21.9	38.3	1.59	+12
25:75	10.55	18.3	37.8	1.63	+ 8
Level of nutrition					
High	—	—	37.0	—	+50
Medium	—	—	37.9	—	− 3
Low	—	—	34.8	—	−38
Production capacity of cows					
High	—	—	39.5	—	− 5
Medium	—	—	35.5	—	− 2
Low	—	—	34.8	—	+12

[a] Scandinavian feed units.

Table 36 Average feed intake and production of cows fed hay and silage plus grain during 189 day feeding trial.

Comparison	Feed units fed	Ration crude fiber	Average daily milk	Pounds of milk per feed unit	Change in body weight
		%	lbs.	lbs.	lbs.
Forage : Grain Ratio					
70 : 30	10.20[a]	24.9	33.8	1.50	+ 44
55 : 45	10.50	22.0	34.3	1.48	+ 48
40 : 60	10.33	18.9	33.9	1.49	+ 26
25 : 75	10.20	15.3	32.8	1.46	+ 8
Level of nutrition					
High	—	—	36.1	—	+108
Medium	—	—	34.8	—	+ 31
Low	—	—	30.4	—	− 45
Production capacity of cows					
High	—	—	38.6	—	+ 24
Medium	—	—	32.8	—	+ 37
Low	—	—	30.0	—	+ 33

[a] Scandinavian feed units.

were fed, the high producing cows were better able to express their superiority than they were on the poorer rations.

The plane of nutrition effects were in keeping with the results of many such trials. With the all hay plus grain rations the medium level of feeding was superior but the high level was superior with the better feeding regime. The data illustrate a commonly observed factor: good rations tend to improve milk production and the response to level of feeding is greater than with poor rations which have no positive influence on milk production. It should also be pointed out that the data do indicate that the hay + grain rations may have been overevaluated by the feeding systems since the cows lost greater quantities of body weight on this ration.

The response of the cows to the four ratios of forage to grain also illustrated the difference between the two feeding regimes as well as the ratios. Top milk production was obtained with the 40:60 hay to grain ratio but with the 55:45 hay + silage + grain ratio. The difference between the two feeding regimes was apparently due to the lower level of crude fiber in the regime which included silage and its influence on feed utilization.

The data illustrate the factors necessary for high levels of efficient milk production: (1) cows with the genetic potential for high levels of production, (2) the use of feeds suited to the production of milk, (3) cows being fed sufficient quantities of rations properly balanced to maintain the type of rumen fermentation necessary to provide the udder with a balanced supply of precursors for efficient synthesis of the components of milk.

WHAT IS MILK PRODUCTION?

On the farm, milk production is measured as the amount of liquid that is sold as milk. In nutrition, milk is a considerably more complex product. To understand today's rations for dairy cows, it is necessary that we first understand milk production as it occurs and thus understand the needs of the udder for milk synthesis and the importance of the ration fed the cow in meeting these needs.

The cow does not manufacture a single substance called milk. She manufactures or synthesizes milk fat, milk protein, and milk sugar. These three ingredients are suspended in water along with minerals and vitamins from the blood stream and the finished product is removed from the udder as milk. An important item in this synthesis is that each of the items is produced largely independent of the others. In short, milk may contain a low percentage of fat without affecting its content of protein or sugar.

Milk Fat is a combination of fatty acids with chain links from C_4 through C_{20}. The primary precursor for milk fat is acetic acid which is formed in the rumen primarily from the fermenation of crude fiber. Acetic acid is the primary source of carbon for the fatty acids C_4 through C_{16}. Butyric acid which also originates in rumen fermentation may be used to form the C_4 through C_{10} acids. The fatty acids C_{16} and above are apparently formed from plasma triglycerides or body fat. The dependence of milk fat formation on acetic acid explains part of the necessity for concern for crude fiber in the ration and the detrimental effects of rations containing too much starch. The ability of the udder to use butyric acid to form part of the milk fatty acids may explain why certain feedstuffs such as barley do not reduce fat percentage when included in the ration at a level that would reduce fat percentage if the feed was corn. This is an example of utilizing feedstuffs to do specific jobs in the ration. If the basic ration consists of lush forage or corn silage with a very high ratio of grain to forage, adding corn in the grain ration obviously aggravates the problem and calls for a grain ration that contains feedstuffs such as barley, brewers grains, citrus pulp, or beet pulp. The importance of plasma triglycerides in the formation of the long-chain acids explains the role of the body condition of the cow in maintaining fat percentage. Since the C_{16} and higher fatty acids constitute only a small portion of the total fat, fat percentage does normally vary greatly due to body condition but it is a factor that must be considered. Unfortunately, there is considerable antagonism between the acetic acid production and utilization and the propionic acid production and utilization that will be seen as a necessary ingredient for the formation of milk lactose. Thus the molar percentage of acetic acid produced in the rumen is reduced when large amounts of starch are fed and the increased propionic acid produced further appears to interfere with the uptake of acetic acid in the blood stream. This unfortunate conflict between the primary precursors of two of the important ingredients of milk may explain another part of the necessity for maintaining a proper balance between the fiber and starch content of rations for dairy cows.

Milk Sugar The primary source of energy for the formation of lactose is blood glucose. Quantitatively the udder removes more glucose for the formation of milk than any other precursor. Linzell, working with lactating goats, showed that the udder removed on the average 10.3 g glucose, 3.2 g of acetic acid, and at least 650 mg. of amino acid nitrogen from the blood for each 100 ml of milk produced. Some estimates indicate that as much as 80 per cent of the carbon for lactose comes from the plasma glucose or from other related sources such as bicarbonate, acetate and propionate. From a feed point of view, the most direct source of glucose would be the propionic acid produced in the rumen and converted to glucose in the liver. Since the glucose requirement is the greatest need of the udder, it is unfortunate that propionic acid cannot be held at a high level in the rumen. The antagonism between the production and utilization of acetic and propionic acids and the general low blood content of glucose results in the glucose supply of the ruminant being a constant critical factor in milk production. The critical balance of glucose becomes of considerable importance in high producing cows when ketosis enters the picture in the early stage of lactation. One of the fertile fields for further research in dairy cattle nutrition is in the area of increasing the supply of glucose to the udder without reducing the needed acetic acid.

Milk Protein The protein content of milk is apparently synthesized from the amino acids in the blood and protein content is largely dependent upon an adequate supply of nitrogen to the cow. There is some tendency for the protein content of milk to be depressed by rations supplying inadequate energy as well as rations containing too high a level of grain for normal milk production.

Minerals and Vitamins Within broad limits, the content of minerals and vitamins reflect the status of these substances in the blood stream. The mineral content of the milk remains relatively constant so long as the cow does not have a mineral deficiency. The content of vitamins, particularly of carotein or vitamin A, is closely related to the vitamin content of the feed. The water soluble vitamins are somewhat more stable since they are produced in rumen fermentation. The fat soluble vitamins—A, D, E—are, of course, higher in milk of higher percentage of butterfat. In essence, milk seems to be one means of secreting the vitamins which the cow has in excess of her needs and, as such, the vitamin content represents the excess of vitamins in the blood stream.

Total Milk As pointed out above, the total milk which can be removed from the cow represents the sum of the synthesis of fat, sugar, and protein and the content of vitamins and minerals all of which have been suspended in water. At the moment, the amount of water used in the suspension appears to be most closely related to the amount of lactose produced. The mineral content obviously influences the amount of water needed to maintain the milk in the udder in an isotonic state with the blood stream, but the total milk is primarily a function of the amount of lactose synthesized in the udder. Total milk production, therefore, may be largely independent of the content of butterfat or milk protein. This is reflected in the total solids content of the milk and explains the common field observation that cows on lush grass or similar feed may produce increased quantities of milk but at the same time produce milk that is sub-standard in butter fat, solids-not-fat, or total solids. The most efficient conversion of feed to milk and the conditions favorable to the production of milk with a normal content of all fractions for the breed of cows being handled is apparently best assured by the feeding of rations containing an optimum content of protein, fiber, and energy. The data in table 37 illustrate the results of an excellent piece of research at the California station comparing rations with a range of grain content from 10 per cent to 100 per cent. The data illustrate the several influences discussed above. The ration containing only 10 per cent grain resulted in the production of a normal percentage of milk fat, a reduced production of lactose and total milk, and a low content of solids-not-fat. This response was rather typical of cows being underfed on energy and also being fed a ration that probably did not support a sufficient production of propionic acid in the rumen. The opposite ration containing 100 per cent grain and a crude

Table 37 The influence of ratio of grain to hay on milk production and milk composition.[a]

Category	Hay : concentrate ratio			
	90:10	60:40	30:70	0:100
Average milk yield, 1 week before trial (lb/day)	50.0	52.9	47.8	50.0
Average milk yield during trial (lb/day)	31.2	38.6	37.7	38.5
Average milk fat percent	3.6	3.6	3.5	2.4
Average FCM (lb/day)	30.0	35.2	34.1	29.8
Average protein in milk (%)	3.46	3.52	3.60	3.46
Average lactose in milk (%)	4.63	4.86	4.57	4.71
Average solids-not-fat (%)	8.80	9.18	9.13	8.91
Maximum intake of dry matter (lb/day)	43.2	41.4	37.7	32.1
Crude fiber of ration (%)	28.7	21.8	14.9	7.3

[a] J. Dairy Science, vol. 49. 1966.

fiber percentage of 7.3 resulted in an increase in lactose and total milk output, no change in protein, a small increase in solids-not-fat, but a drastic reduction in butter fat percentage. The data clearly illustrate the curvilinear nature of the milk production response to ration composition. Although the data do not include information on the rumen production of short chain fatty acids, one can theorize from similar research that the rations created vast differences in the fermentation patterns in the rumen. Until further research clearly outlines all of the relationships between rumen fermentation and milk production, it is suggested that maximum efficiency of feed conversion to milk will take place when rumen fluid contains the following molar composition of volatile fatty acids: Acetic acid, 60 to 65 per cent, propionic acid, 18 to 20 per cent, butyric acid, about 10 to 15 per cent, higher acids, unknown. To maintain the production of rumen volatile fatty acids within these ranges it will be necessary to balance ration with full consideration for their content of protein, energy, fiber and minerals.

FACTORS AFFECTING FEED INTAKE

In today's high-producing dairy herds, the most frequently encountered problem in nutrition is a lack of energy due to an inability of the cow to consume enough feed to meet the requirements of high levels of milk

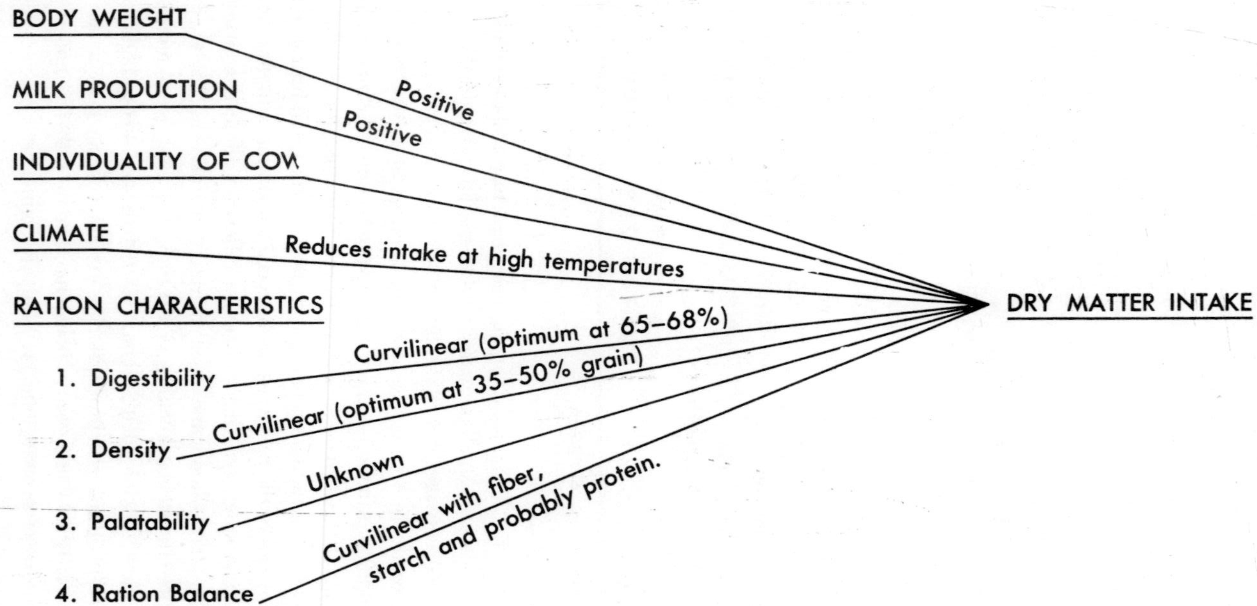

Figure 28 The Influence of Several Animal and Ration Characteristics on Dry Matter Intake by Dairy Cows.

production. For this reason, considerable attention has been given to the factors affecting intake during the last several years. It is unfortunate that the cow does not increase her intake of dry matter with increasing levels of grain feeding and also unfortunate that high energy substances such as fat cannot be efficiently utilized by the cow since this would offer almost endless possibilities for feeding high energy rations. As we have seen throughout this book, the cow maximizes her intake with a combination of forage and grain, thus setting practical limits to her intake from a nutritional point of view. The major factors influencing feed intake are illustrated in figure 28.

Body Weight The positive influence of body weight on intake is easily understood since the larger cow has both a greater capacity and a greater need for energy. As will be seen in the research outlined later, the influence of body weight on intake is, on the average, sufficient to meet the extra requirements for maintenance.

Milk Production There seems to be little doubt that high levels of milk production are accompanied by great appetites. It is also true, however, that the variation from cow to cow in this regard is very great. Not all cows that have the ability to produce high levels of milk are also endowed with great appetites. The result is that some cows produce milk with great losses of body weight while others are able to more nearly meet their energy requirements with increased intake. On the average, the increase in intake with increasing levels of milk production is sufficient to meet the requirements for energy up to the ceiling imposed by the capacity of the cow. In other words, most high producing cows are unable to consume enough feed in early lactation to meet their needs for both milk production and the maintenance of body weight, but their appetite is at its maximum and their level of intake is considerably greater than during their dry period. Within a herd, those cows producing the highest levels of milk at a given time will, on the average, also have the greatest appetites. Another factor that aids in this relationship is the general ability of high producing cows to improve the efficiency with which they convert feed to milk. This ability permits a greater output of milk with a smaller intake of feed.

Individuality of the Cow This factor is well known to men who have fed cows in stanchion barns and have thus had an opportunity to know the individual tastes and preferences of each cow. In the hands of an excellent observer, this knowledge can add many pounds of milk to the lactation average of a cow. Unfortunately, in today's large herds and with the necessity for mass handling and feeding techniques, this factor must now be considered as a negative value in a cow and cows with individual preferences for feeds that are sufficiently different from the average of the herd will be culled for low milk production.

Ration Characteristics The factors which can be built into the ration which will influence intake are of importance since these are the primary points of control that can be exercised in the feeding program.

1. Digestibility The digestibility of the ration exerts its primary influence through the rate of passage of the feed through the digestive tract and the percentage of nutrient availability of the ration. As was pointed out in an earlier chapter, this relationship between intake and digestibility is a curvilinear one having a maximum in most rations between 65 and 68 percent digestibility. Thus both high and low percentages of digestibility exert adverse influences on feed intake.

2. Density The density (weight per volume) of rations apparently also exerts a curvilinear influence on feed intake. This area of investigation is too new to have rations balanced for this characteristic, but future developments will bear considerable watching and new information should be incorporated into ration balancing as soon as possible. In most dairy rations, the major influences of density can be overcome by including 35 to 55 per cent grain in the total ration.

3. Palatability This term has been much used over the years but its usefulness has probably been overemphasized. There is no reason to think that cows do not find some rations more palatable than others or even that individual cows do not have preferences for feed. Since cows cannot communicate with humans, there seems to be little reason to believe that anything other than extreme cases of low palatability can be considered in feeding dairy cows. It is important that palatability of the ration not be confused with changes in ration digestibility, ration density or ration balance. The fact that cows refuse part or all of a ration when first offered a new feed does not constitute a useful measure of palatability since they may consume the new ration in large quantities after a few days time. By the same token, offering a cow a variety of feeds and measuring palatability on the basis of the one most frequently selected is no measure of the appetite the cow may have for an individual feed when it is offered alone. There is little real evidence to support any real concern for palatability over a wide range of feeds that may be included in dairy rations so long as they are free of mold, poisons, or items of known low palatability and so long as the cow is offered an otherwise well balanced ration.

4. Ration Balance Feed intake is influenced by ration balance primarily in relation to the influence of the ration on rumen fermentation. Thus ration balance for crude fiber, starch level and other factors which may slow up passage of the feed or improve rumen fermentation is important in feed intake. A frequently quoted example is that of protein. Rations for some animal functions such as maintenance can be calculated in which the protein requirement for body functions is lower than the protein requirement for good rumen fermentation. For this reason, no matter how low the calculated protein requirement may be, it is not recommended that less than 10 per cent protein be included in the ration to insure good rumen fermentation and feed intake.

CALCULATING FEED INTAKE

Although feed intake is the most important single item in balancing rations for dairy cows, none of the presently used feeding standards contain any real consideration of this item. Most methods for developing rations for cows either assume a constant level of intake or an ability on the part of the cow to consume whatever feed may be needed to carry the recommended levels of nutrients when rations contain a percentage of each nutrient. Our proposal here is to introduce calculated feed intakes into the program so that nutrient requirements can be stated as percentages of the dry matter content of the rations. There are two primary sources of data for making such calculations and we will consider them at this point.

Ohio Data In 1964, Conrad and his co-workers at the Ohio experiment Station published the results of their research on the factors which influence the dry matter intake of dairy cows. The work will undoubtedly become one of the classics in the field of animal nutrition. The basic equation in the findings was:

$$\text{Total dry matter intake} = \frac{\text{Weight}}{1000} \cdot \frac{10.7}{1 - \%\ \text{digested}}$$

The equation says in effect that cows on average dairy rations excrete 10.7 lbs. of indigestible dry matter per 1000 pounds of body weight per 24 hours. Using this value as the base value, a major portion of the variability among animals was accounted for when requirements for maintenance and milk production were added to the equation. When this was done, the equation for calculating maximum feed intake was:

Maximum feed intake

$$= 10.7 \times \frac{W}{1000} + .058\ W^{.73} + .33\ \text{Milk} + .53$$

The equation now contains a factor for the daily output of indigestible residue, a factor for maintenance requirements and a factor for milk production along with the constant that all regression equations contain—in this case .53. Using this data, another equation can be formed to calculate the minimum digestibility that the ration must have to permit the cow to consume the amount of feed calculated as her maximum feed intake. The equation for minimum allowable digestibility is:

$$\text{Minimum allowable digestibility} = \frac{.058\ W^{.73} + .33\ M + .53}{\text{Maximum feed intake}}$$

The two equations have been used in Georgia over the last four years to calculate the rations to be used on hundreds of dairy farms with excellent results. This work on dairy farms and the success that has been registered in improving milk production under actual farm conditions provides the confidence needed to make the formulas the basis of the feeding allowances to be recommended later in this chapter.

The question of the relative accuracy of the calculated values for feed intake is a very real one. The calculated feed intakes appear to be within 5 per cent of the measured intakes in most herds. Since this is true, the suggested ration can be adjusted to the actual intake of the herd to which it is being fed providing the ration has been balanced to provide for such adjustment without significantly altering the balance of nutrients being supplied and a safety factor is included to provide for adequate intake of nutrients at the lower level of intake.

Georgia Data The Georgia data originated from a statistical study of the results of 39 feeding trials conducted at several different experiment stations throughout the United States. The trials included cows with body weights from 1097 to 1569 pounds and levels of milk production from 24 to 70 pounds. The data used were from trials that met the following basic

Table 38 Influence of several factors on ENE intake.[a]

Variable	ENE intake (T) required	ENE intake (T) found	Percent of total
Increase/100 lb. body weight	.63	.969	60
Increase/lb. 4% milk	.30	.601	20
Increase/lb. weight gain	—	4.774	4
Increase/% grain in ration	—	.009	4
Not accounted for	—	—	16

[a] Based on 39 feeding trials. Range of body weight 1097 to 1569 lbs. Range of milk production 24 to 70 lbs.

requirements: (1) cows were fed free choice rations, (2) definite ratios of grain to forage were fed all cows, and (3) data on the digestibility of the rations were available. The best-fit equation describing the factors influencing dry matter intake was:

$$Y = 5.380 + .008\ BW + .359\ M + 4.719\ WG - .028 \text{ per cent grain in ration}$$

The equation accounted for 87 per cent of the variation in observed intake and the standard error was about 7 per cent of the mean. The equation would therefore be expected to predict the intake of the ration within 7 per cent of the actual intake. The data in table 38 indicate that the cows increased their intake over the average intake of the group when their body weight was above average or their milk production was above the group average.

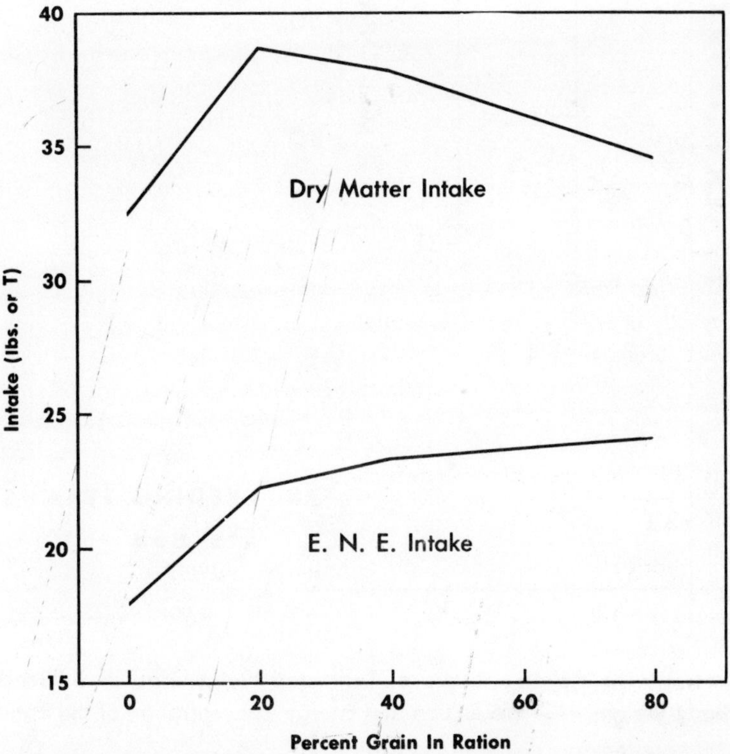

Figure 29 The Relationship Between Percent Grain in the Ration and the Intake of Dry Matter and Estimated Net Energy.

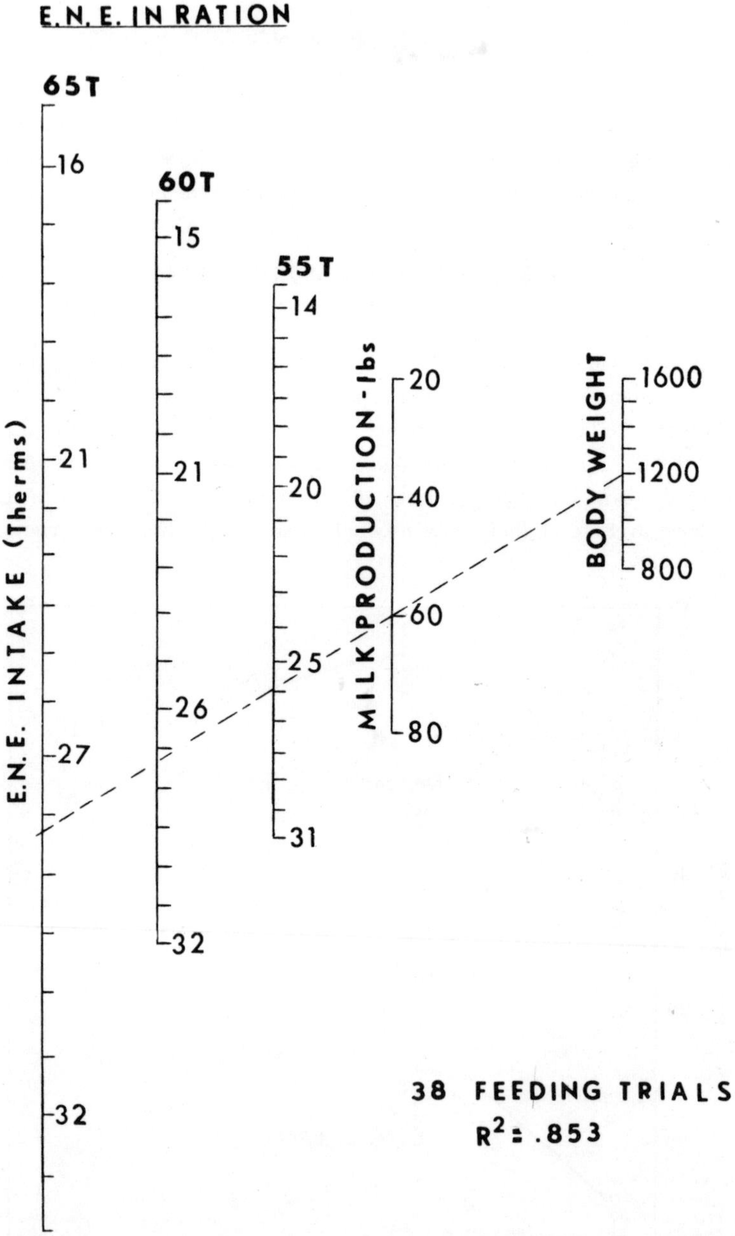

Figure 30 The Relative Intake of Estimated Net Energy as Influenced by Body Weight, Milk Production and Energy Concentration of the Ration.

The comparison between the magnitude of these increases and the requirements for maintenance and milk production indicated that the increased intake was sufficient to meet the requirements of the cows. This data is basic for recommending the feeding of a single ration to groups of cows with the idea that cows with differences in body weight or level of milk production will be able to compensate for their requirements with increases in feed consumption.

The data in figure 29 illustrate the general relationships between the grain to forage ratio in the ration and intake of total dry matter and estimated net energy. The data clearly illustrate that grain to forage ratio has a curvilinear effect on dry matter intake with an optimum somewhere around 40 per cent grain in the ration. The variation encountered between rations indicated that this optimum area probably extends from 35 to 55 per cent grain in the ration with a decreased intake at either a lower or higher percentage of grain being fed.

In figure 30, the data from the Georgia research have been transferred to a chart from which a calculated intake of estimated net energy can be determined for individual cows or groups of cows when the body weight, level of milk production and ENE content of the ration are known.

The data in table 39 compare the Ohio and Georgia system for calculating feed intake in terms of three levels of milk production and three body weights. It is obvious that the systems are nearly identical for the three levels of milk production with the cows weighing 1000 pounds. The Ohio system gives

Table 39 **The calculated intake of dry matter by cows at three body sizes and three levels of milk production by the formulas suggested by research in Ohio and Georgia.**

Milk production	Dry matter intake (lbs. per day)			
	Body weight	1000	1200	1400
lbs.				
40	Ohio	33.4	36.8	40.2
	Georgia	31.7	33.2	34.8
60	Ohio	40.0	43.4	46.8
	Georgia	39.0	40.6	42.4
80	Ohio	46.6	50.0	53.4
	Georgia	46.0	47.6	49.2

Table 40 The calculated daily TDN requirements of six cows being fed 25 lbs. alfalfa hay and the calculated requirement for grain to meet the cow's TDN needs.

Cow No.	Body weight	Daily milk production	TDN required	TDN from 25 lb. hay	Pounds of grain required	Total feed
	lbs.	lbs.	lbs.	lbs.	lbs.	lbs.
			Calculated Requirements			
1	1340	52	27.4	13.7	20.0	45
2	1310	26	18.8	13.7	7.3	33
3	1000	20	14.4	13.7	1.0	26
4	1200	65	29.6	13.7	22.7	48
5	1200	42	23.0	13.7	13.2	38
6	1000	46	22.7	13.7	12.9	38

Alfalfa hay = 55% TDN & 16% protein.
Grain ration = 70% TDN & 12% protein.

a greater estimated intake with larger cows which probably reflects the influence of the constant in the Ohio equation for indigestible residue. Although either of the systems could be used for the basis of recommended feeding standards or programs, the Ohio system will be used in the calculation of percentage of nutrients for optimum rations.

Table 41 Some characteristics of the rations calculated for the six cows.

Cow No.	Feed/100 lb. body wt.	Percent in ration: protein	Percent in ration: TDN	Calculated maximum intake	Minimum digestibility
	lbs.	%	%	lbs.	%
1	3.35	14	62	43	66
2	2.52	15	57	34	59
3	2.60	16	55	27	62
4	4.36	14	62	43	73
5	3.17	15	60	37	67
6	3.80	15	60	35	71

OPTIMUM RATIONS—WHY AND HOW?

With the many systems for rationing feed to dairy cows now in use throughout the world, when one proposes a new program it would appear necessary to explain in some detail why the program is needed. The primary program in use in the United States is the so-called TDN system which uses tables of feed composition or the results of "forage testing" and tables of nutrient requirements to formulate rations for either individual cows or herds. The system has been modified in Dairy Herd Improvement work to calculate the requirements and rations on a calculated net energy basis but retains the general principles of the TDN system. Great credit must be given the system despite all of its shortcomings since many herds are better fed with the system than without any system at all. The primary problems of the system are illustrated in the calculations in tables 40 and 41. The data illustrate the results of using the TDN system for calculating the amounts of grain to be fed individual cows when they are fed a basal ration of 25 pounds of good alfalfa hay per day and enough grain to meet their TDN requirements. The ration selected is basically a good milking ration and the results from the calculations are not as bad from a nutritional point-of-view as could have been illustrated had a low quality forage been selected.

The 25 pounds of alfalfa hay was calculated to supply 13.7 (55 per cent TDN) pounds of TDN and 16 per cent protein to the ration. The grain ration indicated was a good 70 per cent TDN grain mix having 12 per cent protein. The calculations in table 40 are straight-forward values in which the 13.7 pounds of TDN supplied by the hay were subtracted from the total TDN required and the difference to be supplied by the grain was calculated by dividing the pounds of TDN to be supplied by the grain by 70 per cent.

The data in table 41 evaluate the rations calculated for each cow in terms of modern knowledge in dairy cattle feeding. Although both the hay and the grain ration was from high quality feeds, it will be seen that the one common deficiency of all the rations was a failure of the rations to be up to the minimum digestibility calculated to be required to insure the needed level of intake. In the case of cows 4 and 6, the amount of feed required was probably in excess of their capacity and the cows would either consume all of the hay and leave grain or the reverse. In either case, the rations actually consumed would have been considerably different in composition than the rations as calculated. Only cow 2 with the lowest level of milk production would have been receiving a ration that was close to her needs. Unless this cow was rationed by individual feeding, the amount of feed calculated to meet her needs was below her expected appetite and the cow would probably have consumed excess feed. It can also be speculated that the rations as calculated would have resulted in declines in milk production in the lower producing cows due to the feeding of a low energy ration and to declines in milk production in the high producing cows due to too little energy in the total diet unless the cows were in early lactation and would lose weight to make milk. In addition to the obvious deficiencies in the system for rationing cows, the suggested rations do not consider other factors such as decline in feeding value of rations due to plane of feeding and the need to consider the influence of the rations on rumen fermentation.

The deficiencies of the conventional method for rationing dairy cows have been recognized for many years and many alternatives have been suggested. One of the simpler and more workable programs was the one suggested by Dr. J. T. Reid (J. Dairy Sci. 44) some years ago. In essence, his program was the same as the "challenge feeding" idea that received considerable attention in the early 1960's. Reid's plan consists of: "(a) The ad libitum feeding of the best quality forages available. (b) Flexible energy allowances during the last two months of the dry period which, in some instances and depending on the condition of the cow, would be provided by a ration ranging from all forage to forage plus 8 to 10 lbs. of concentrates during the first 4 to 6 wks., and a gradually increasing level of concentrates thereafter, reaching a level of 15 to 18 lbs. by the time of calving. This scheme would appear to have two benefits: the storage of body tissue and, perhaps more important, accustoming the cow to consume large quantities of feed. (c) After calving, increasing the level of concentrates as rapidly as possible to either maximum appetite or maximum milk yield, whichever is attained first. The cow should decide her level of intake at this time; the dairyman must learn by trial the rate at which the individual cow's intake can be increased. (d) Thereafter, the level of concentrates should be reduced to the lowest level which does not reduce the milk yield. In this way feed intake tends to lead the milk output, rather than the reverse."

Many dairymen adopted the basic principles of Reid's plan and many herds responded with spectacular results. The amount and nature of the milk response was of course in direct proportion to the level of feeding that had been practiced in the herd and the milk producing capacity of the herd. The primary problem with this method of feeding dairy cows was recognized by Reid in his paper: "Such a feeding plan, however, puts more daily management pressure on the dairyman and requires that each cow be given individual attention." Across the world there are many cows that have the good fortune to be handled in small herds with the skilled individual attention required to make this program work. Where this is true, there is probably no finer method for feeding dairy cows. In other herds, lacking either the time for individual attention or the skill for such a feeding program, the system is capable of creating great problems since all cows tend to be overfed on grain, underfed on forage and the whole program poorly carried out. It is to such herds—large herds fed in groups or as a single group, milked in milking parlors without daily milk weights and handled with maximum emphasis on the individual cow taking care of herself—that the idea of "optimum rations" is directed.

RECOMMENDED NUTRIENT ALLOWANCES

To establish the requirements for optimum rations it was necessary to first establish tables of nutrient allowances based on the requirements for energy, protein and minerals. The results of this effort are presented in the five tables which follow.

Energy Requirements The recommended requirements for metabolizable energy are in table 42. The recommendations were derived by bringing into play all of the factors known to affect this area of nutrition, some of which are deserving of considerable explanation and discussion.

In all cases, the requirements have been calculated in terms of a ration containing 2.6 Mcal/kg of metabolizable energy. This ration would have a dry matter digestibility in the range of 65 to 68 per cent and a net energy content of 56 to 60 Therms. This ration was used since it meets the requirements of a ration suitable for use in producing high levels of milk production based on the need for a high level of dry matter digestibility to insure intake and also on the need for a ration that is optimum for efficient conversion of feed to milk.

The first step in preparing the table was to calculate the requirements for maintenance and milk production for each level of body weight and milk production. This was done by assuming a requirement for maintenance of 131 kcal m.e. per $kg.^{75}$ and a requirement of 1072 kcal/kg of 4% milk. The sum of the two estimates of required metabolizable energy for each level of weight and milk production is shown in the column headed M. E. (metabolizable energy). The next step was to calculate the amount of dry matter required to provide this amount of metabolizable energy. To do this, it was necessary to correct the value of 2.6 Mcal/kg for the loss of energy due to level of intake. The corrections used were those recommended in the Agricultural Research Council's Technical Reviews and Summaries. This correction is very important since it accounts for the loss of digestibility with high rates of feeding and also corrects for the loss in efficiency of feed utilization with increasing levels of feeding. The dry matter required values are listed in the column headed "dry M" and are the result of the equation.

$$\text{D. M. Req.} = \frac{\text{metabolizable energy required}}{\text{metabolizable energy}}$$
$$\text{(corrected for feeding level)}$$

The third column in each group is the CMI (calculated maximum intake) of dry matter based on the results of the Ohio State research. This value is included as a measure of the relative ability of the cow to consume the required amount of feed needed to supply her energy requirements. The relative ability of the cows to consume dry matter can only be estimated and will vary from herd to herd, but by comparing the amount of dry matter required with the calculated maximum intake a good estimate of the level of milk production which can be supported without loss of body weight is provided. If it is assumed that cows in early stages of lactation actually are more efficient users of feed, then by selecting the first level of milk which indicates that expected intake is below required intake as the break-even point, we would estimate that a 900 pound cow should produce about 50 pounds of milk without loss of weight and that this value increases with body size to 60 plus pounds for a cow weighing 1300 pounds. The value of this estimate lies in its usefulness in estimating the level of milk that can be supported by feed intake and thus the point at which special supplementation may necessary. In preparing the tables for protein and minerals, the calculated maximum intake value was used. The fact that the cow would be expected to be unable to consume enough energy for top levels of milk production should not preclude her consumption of sufficient levels of other nutrients providing they are included in the ration on the basis of her expected level of intake.

Digestible Protein Requirements The requirements for digestible protein are a combination of the recommended requirement for maintenance suggested by the British Agricultural Research Council and the requirements for milk suggested by Reid, et al. (J. Dairy Science, 49). As in all such calculations, the requirements used are those which the author feels represent the best estimates based on research results and the underlying principles of nutrition.

Since protein is usually calculated in terms of crude protein and not digestible protein, it is necessary to keep in mind the approximate crude protein levels which correspond to the desired levels of digestible protein.

They are:

Crude protein		Digestible Protein
16%	=	11.6%
14	=	9.7
12	=	7.8
10	=	5.9

Based on the above calculations, the crude protein required to meet the cow's digestible protein varies from 10 to 16 per cent of the total ration. Rations containing 14 per cent crude protein would meet most of the levels of production in which the cow would also be expected to consume sufficient energy to meet her needs. The data to indicate that cows yielding levels of milk in excess of their ability to consume energy may need special attention to their protein needs. In most herds, formulating rations with 12 to 14 per cent crude protein would cover the need of the majority of the cows. The data do indicate that it may be advisable to consider providing a special high protein supplement for cows producing in excess of 5 pounds of 4 per cent milk per 100 pounds of body weight.

Mineral Requirements The mineral requirements were calculated on the basis of the recommendations of the Agricultural Research Council and the percentage requirement on the basis of calculated maximum intake. The requirements for calcium increased quite rapidly with increasing levels of milk production which is in keeping with the well known high level of calcium in milk. The magnesium requirement was little affected by production and the phosphorus requirement was intermediate. On the basis of these calculations, there appears to be little reason for changing the frequently recommended requirements of .7 per cent calcium, .5 per cent phosphorus and .15 per cent magnesium in the total ration for milk production.

Vitamin Requirements The only known vitamin requirements of dairy cows are for the fat soluble vitamins A, D, and E. While there is a great volume of research data on the requirements of these vitamins, they are so widespread in the natural feedstuffs of ruminants that it is difficult to justify any consideration for adding them to rations for dairy cows. Since all three vitamins are available in low-cost supplements, many people prefer to routinely add them to the ration as insurance against even the low probability of a deficiency. Since the toxic level of the three vitamins is several thousand times their requirements, there appears to be no nutritional reason why such insurance supplementation should not be done. No exact level of supplementation can be given and the usual recommendation is to supplement according to personal desires. A frequently used level is 2,000,000 units of vitamin A, 1,200,000 units of D and 60,000 units of vitamin E per ton of grain ration.

CONCLUSIONS

Utilizing all of the data on nutrient requirements and the expected levels of intake for cows of several body sizes and levels of milk production, the recommended composition of optimum rations to be fed free choice can be listed as follows:

1. The ration should contain between 35 and 55 per cent grain to insure maximum intake in terms of ration density.

2. The ration should contain 2.6 Mcal/kg of metabolizable energy or its equivalent of 65 to 68 per cent digestibility or 56 to 60 Therms of net energy per pound.

3. The ration should contain 18 to 20 per cent crude fiber if the grain and forage are fed separately or 16 to 20 per cent crude fiber if the grain and forage are mixed and fed as a single ration.

4. The ration should contain 12 to 14 per cent crude protein depending upon the level of milk production of the herd. Consideration should be given to providing additional protein for cows producing in excess of 5 pounds of 4 per cent milk per 100 pounds of body weight.

5. The ration should contain 0.7 per cent calcium, 0.5 per cent phosphorus and 0.15 per cent magnesium. Unless there is a specific trace mineral deficiency in the locally grown feeds, there is no evidence of a need for other mineral supplementation. Salt may be added to the ration or fed free choice.

6. Vitamins A, D, and E may be added if desired, but their need is difficult to demonstrate when cows are fed the quantity and quality of forage that will be required to permit high levels of milk production.

References

1. Agricultural Research Council, The Nutrient Requirements of Farm Livestock. No. 2—Ruminants. Technical Reviews and Summaries. London 1965.

2. Conrad, H. R., Pratt, A. D., and Hibbs, J. W. Regulation of Feed Intake in Dairy Cows. I. Change in Importance of Physical and Physiological Factors with Increasing Digestibility. J. Dairy Science 47:54. 1964.

3. Nordfeldt, Sam and Ruudvere, Aksel. Influence of Various Levels of Roughage

Table 42 Approximate levels of dry matter intake required, metabolizable energy required and calculated maximum dry matter intake at several levels of 4% milk production and body weight by cows fed a balanced ration containing 2.6 Mcal/kg dry matter at maintenance.

Level of milk	Body weight:	900			1000			1100			1200			1300		
	Item:	Dry M	M.E.	CMI	Dry M	M.E.	CMI	Dry M	M.E.	CMI	Dry M	M.E.	CMI	Dry M	M.E.	CMI
lbs.		lb.	Mcal	lb.	lb.	Mcal	lb.	lb.	Mcal	lb.	lb.	Mcal	lb.	lb.	Mcal	lb.
20		22.6	26.0	25.0	23.2	26.7	26.8	25.1	28.9	28.5	26.3	30.2	30.2	27.3	31.4	31.9
30		27.5	30.8	28.3	28.2	31.6	30.1	29.8	33.7	31.8	30.9	35.0	33.5	32.1	36.3	35.2
40		31.9	35.7	31.6	33.4	36.4	33.4	34.5	38.6	35.1	35.6	39.8	36.8	36.6	41.0	38.5
50		37.2	40.6	34.9	37.9	41.3	36.7	39.5	43.5	38.4	41.1	44.8	40.1	41.1	46.0	41.8
60		41.6	45.4	38.2	43.5	46.1	40.0	44.4	48.3	41.7	45.5	49.6	43.4	46.6	50.8	45.1
70		47.5	50.3	41.5	48.2	51.0	43.3	49.5	53.2	45.0	51.0	54.5	46.7	51.0	55.6	48.4
80		53.6	55.2	44.8	54.4	56.0	46.6	54.9	58.1	48.3	56.0	59.3	50.0	57.2	60.6	51.7

$$\text{Dry matter required} = \frac{\text{Met. energy required}}{\text{Met. energy per lb (corrected for level of intake)}}$$

Metabolizable energy required = maintenance (131 kcal/kg$^{.75}$) + production (1072 kcal/kg 4% milk)

$$\text{Calculated maximum intake} = 10.7 \times \frac{W}{1000} + .058\ W^{.73} + .33M + .53$$

Where = W = body weight (pounds)
M = 4% milk

Alternate formula = $Y = 5.380 + .008X_1 + .359X_2 + 4.719X_3 - .028X_4$ ($R^2 = .866$)
Y = d.m. intake (kg)
X_1 = body weight (kg)
X_2 = milk production (kg)
X_3 = weight gain (kg)
X_4 = per cent grain

Table 43 Approximate daily requirements of digestible protein, and percentage of total dry matter intake for cows being fed a balanced ration containing 2.6 Mcal/kg metabolizable energy.

Level of milk	Body weight: 900 (408kg)		1000 (454)		1100 (498)		1200 (544)		1300 (590)	
	Item: g/day	% of ration	g/day	% of ration	g/day	% of ration	g/day	% of ration	g/day	% of ration
20	733	6.4	755	6.2	771	5.9	792	5.8	812	5.6
30	1002	7.8	1024	7.5	1040	7.2	1061	7.0	1081	6.8
40	1279	8.9	1301	8.6	1317	8.3	1338	8.0	1358	7.8
50	1569	9.9	1591	9.5	1607	9.2	1628	8.9	1648	8.7
60	1869	10.8	1891	10.4	1907	10.1	1928	9.8	1948	9.5
70	2179	11.6	2201	11.2	2217	10.8	2238	10.5	2258	10.3
80	2507	12.3	2529	12.0	2545	11.6	2566	11.3	2586	11.0

Maintenance— 900—219g
1000—241g
1100—257g
1200—278g
1300—298g

Milk—20 lbs.—25.7 g/lb.
30 lbs.—26.1 g/lb.
40 lbs.—26.5 g/lb.
50 lbs.—27.0 g/lb.
60 lbs.—27.5 g/lb.
70 lbs.—28.0 g/lb.
80 lbs.—28.6 g/lb.

Table 44 Approximate daily requirements of calcium, and percentage of total dry matter intake for cows being fed a balanced ration containing 2.6 Mcal/kg metabolizable energy.

Level of milk	Body weight: Item:	900 (408kg) g/day	900 (408kg) % of ration[a]	1000 (454) g/day	1000 (454) % of ration	1100 (498) g/day	1100 (498) % of ration	1200 (544) g/day	1200 (544) % of ration	1300 (590) g/day	1300 (590) % of ration
20 lbs.	(9.07 kg)	40.3	.36	42.0	.35	43.6	.34	45.2	.33	46.9	.32
30	(13.6 kg)	53.2	.41	54.9	.40	56.4	.39	58.1	.38	59.7	.37
40	(18.1 kg)	66.0	.46	67.7	.45	69.3	.43	70.9	.42	72.5	.41
50	(22.7 kg)	79.1	.49	80.8	.48	82.4	.47	84.0	.46	85.6	.45
60	(27.2 kg)	92.0	.53	93.6	.51	95.2	.50	96.8	.49	98.5	.48
70	(31.7 kg)	104.8	.56	106.4	.54	108.0	.53	109.6	.52	111.3	.51
80	(36.3 kg)	118.0	.58	119.6	.56	121.1	.55	122.8	.54	124.4	.53

[a] Based on calculated maximum intake.

Maintenance = 3.55 gm Ca/100 kg.
Milk = 2.85 gm Ca/kg 4% milk.

Table 45 Approximate daily requirements of phosphorus, and percentage of total dry matter intake for cows being fed a balanced ration containing 2.6 Mcal/kg metabolizable energy.

Level of milk	Body weight: / Item:	900 (408kg) g/day	900 (408kg) % of ration[a]	1000 (453) g/day	1000 (453) % of ration	1100 (499) g/day	1100 (499) % of ration	1200 (544) g/day	1200 (544) % of ration	1300 (589) g/day	1300 (589) % of ration
20 lbs.	(9.1 kg)	34.6	.30	37.8	.31	40.9	.32	44.1	.32	47.2	.33
30	(13.6 kg)	42.3	.33	45.5	.33	48.7	.34	51.9	.34	55.0	.34
40	(18.1 kg)	50.1	.35	53.2	.35	56.4	.35	59.6	.36	62.7	.36
50	(22.7 kg)	58.0	.37	61.1	.37	64.3	.37	67.5	.37	70.6	.37
60	(27.2 kg)	65.8	.38	68.9	.38	72.1	.38	75.3	.38	78.4	.38
70	(31.7 kg)	73.5	.39	76.6	.39	79.8	.39	83.0	.39	86.1	.39
80	(36.3 kg)	81.4	.40	84.5	.40	87.7	.40	90.9	.40	94.0	.40

[a] Based on calculated maximum intake.

1.72 g/kg milk
14.9 gm at 350 kg + .07g/kg

Table 46 Approximate daily requirements of magnesium, and percentage of total dry matter intake for cows being fed a balanced ration containing 2.6 Mcal/kg metabolizable energy.

Level of milk	Body weight:	900 (408kg)		1000 (454)		1100 (498)		1200 (544)		1300 (590)	
	Item:	g/day	% of ration	g/day	% of ration	g/day	% of ration	g/day	% of ration	g/day	% of ration
20 lbs.	(9.07 kg)	13.9	.12	14.6	.12	15.3	.12	16.0	.12	16.5	.11
30	(13.6 kg)	16.8	.13	17.5	.13	18.2	.13	18.9	.12	19.4	.12
40	(18.1 kg)	19.6	.14	20.3	.13	21.0	.13	21.7	.13	22.2	.13
50	(22.7 kg)	22.5	.14	23.2	.14	23.9	.14	24.6	.13	25.1	.13
60	(27.2 kg)	25.3	.14	26.0	.14	26.7	.14	27.4	.14	27.9	.14
70	(31.7 kg)	28.2	.15	28.9	.14	29.6	.14	30.3	.14	30.8	.14
80	(36.3 kg)	31.0	.15	31.8	.15	32.5	.15	33.2	.15	33.7	.14

0.63 gm Mg/kg milk
.0152 gm/kg body wt } + 2. g safety

and Concentrate and the Effect of Different Planes of Nutrition on Milk Production in Dairy Cows. Lantbrukshogskolans Annaler 29:345. 1963.

4. Ronning, Magnar and Laben, R. C. Response of Lactating Cows to Free-Choice Feeding of Milled Diets Containing from 10 to 100% Concentrates. J. Dairy Science 49:1080. 1966.

5. Reid, J. T. Problems of Feed Evaluation Related to Feeding Dairy Cows. J. Dairy Science 44:2122. 1961.

6. Reid, J. T., Moe, P. W., and Tyrrell, H. F. Energy and Protein Requirements of Milk Production. J. Dairy Science 49:215. 1966.

Chapter 7

Feed Management to Increase Profit

To include a chapter on farm management in a book devoted to the principles of feeding dairy cattle may require some explanation. It is included because there are certain input-output functions which are found only in milk production. With growing or fattening animals, the level of feeding employed at one time is not likely to exert a detrimental influence on animal response at a later date. It was pointed out in the chapter on growth that a low level of feeding might result in a more efficient use of feed if the animals were later changed to a high plane of nutrition. In milk production a wrong decision in feeding at one stage in lactation usually results in irreparable damage for the remainder of that lactation. In fact, poor feeding practices which fail to bring the cow to a good state of body condition at calving time may reduce total milk production in the subsequent lactation since the cow has a limited store of body reserves to call on in early lactation when production demands exceed appetite potentials.

There are three areas of nutrition which can be manipulated in milk production to influence total income above feed costs. They are: (1) the cost of the feed ingredients used in rations, (2) the quality of the ingredients used in rations, and (3) feeding the proper rations at the proper times.

THE COST OF THE INGREDIENTS USED

Most students of dairy farm management would agree that the best economic measure of the success of the feeding program in milk production is the factor "income above feed cost." This is not to indicate that the nutritional portion of dairy farm management is the only area determining net profits. It is to say that full attention must be afforded the contribution made by the feeding program to net profit.

It may be useful to look at the basic function of the dairy cow in the total farm program. On some farms, it sometimes appears that this basic function has been forgotten. *The dairy cow serves as a manufacturing plant to change low-value or otherwise useless feeds into a high quality foodstuff*

—*thus, increasing the market value of the original feed.* As such, milk production shares with most other manufacturing operations several general characteristics. They are: (1) the use of least-cost raw materials suited to efficient manufacturing processes, (2) maximum efficiency in use of the machinery (cow) which demands high levels of production with minimum maintenance costs, and (3) the production of a product suited to market demands.

The basic conversion of feed for profit can be illustrated in the following example for the conversion of silage to milk. Using values of $8.00/ton for corn silage, $70/ton for grain and $5.50/cwt for milk, let us examine the feeding of one cow for one month. A suitable ration for our cow producing 50 lbs. of milk daily would be:

80 lbs. corn silage
18 lbs. of grain

On a 30-day basis our feed inputs would be:

2400 lbs. silage	=	**$ 9.60**
540 lbs. grain	=	**18.90**
Total	=	**$28.50**

Our 30-day output of milk would be:

1500 lbs. of milk = $82.50

If we subtract the $18.90 we paid for the grain from the $82.50 we received for the milk, we have a gross return for the 2400 lbs. of silage of $63.60 or $53 per ton of silage. This conversion of $8/ton silage represents the basic function of the milk production operation.

In recent years, there has been considerable concern as to whether or not this conversion is most efficient in terms of producing income when feeds are produced or purchased. In tables 47 and 48, the results of extensive studies by Hoglund at Michigan State are presented. Since one of his objectives was to consider future potentials, the results are presented for both a 40 and a 160 cow herd on three land evaluations. The data leave little doubt that the production of at least the forage portion of the ration on the farm will continue to be a profitable part of milk production for many years.

It may appear superfluous to mention the fact, but it is of considerable importance to be concerned with the basic cost of both the forage and the grain portion of the ration. In recent years we have heard much of the so-

Table 47 Calculated receipts, expenses and labor income—Grow versus buy feed for a 40-cow dairy farm—Medium level of prices and good crop management practices.

Item	Grow All Feed		Buy Grain		Buy All Feed	
Lbs. milk sold/cow	12,000	14,000	12,000	14,000	12,000	14,000
	Dollars	Dollars	Dollars	Dollars	Dollars	Dollars
Receipts:						
Milk @ $4.50 cwt.	21,600	25,200	21,600	25,200	21,600	25,200
Cull cows and calves	3,200	3,200	3,200	3,200	3,200	3,200
Total receipts	24,800	28,400	24,800	28,400	24,800	28,400
Expenses:						
Seed, fertilizer & lime	2,112	2,196	1,500	1,500	—	—
Fuel, oil & custom hire	1,176	1,241	733	733	—	—
Depreciation, repairs, insurance						
Buildings & storage	4,196	4,196	4,196	4,196	4,196	4,196
Power & crop machinery	2,699	2,699	2,699	2,699	—	—
Feed (including protein & urea)	1,096	1,340	4,792	5,564	14,884	15,656
Bedding	600	600	600	600	600	600
Labor	1,400	1,400	1,400	1,400	—	—
Miscellaneous	1,520	1,920	1,520	1,920	1,520	1,920
Taxes and interest						
Land/$300 A.	6,025	6,258	5,080	5,200	2,350	2,570
Land/$450 A.	7,442	7,731	5,115	6,235	2,425	2,645
Land/$600/A.	8,860	9,205	7,150	7,270	2,500	2,720
Total expenses—Land/$300 A.	20,823	21,850	22,520	23,812	23,550	24,942
—Land/$450 A.	22,241	23,323	23,555	24,847	23,625	25,017
—Land/$600 A.	23,659	24,797	24,590	25,882	23,700	25,092
Labor Income:						
Land/$300 A.	3,076	6,550	2,280	4,588	1,250	3,458
Land/$450 A.	1,659	5,077	1,245	3,553	1,175	3,383
Land/$600 A.	1,141	3,603	210	2,518	1,100	3,308

Table 48 Calculated receipts, expenses and labor income—Grow versus buy feed for a 160-cow dairy farm—Medium level of prices and good crop management practices.

Item	Grow All Feed		Buy Grain		Buy All Feed	
Lbs. milk sold/cow	12,000 Dollars	14,000 Dollars	12,000 Dollars	14,000 Dollars	12,000 Dollars	14,000 Dollars
Receipts:						
Milk @ $4.50 cwt.	86,400	100,800	86,400	100,800	86,400	100,800
Cull cows and calves	12,800	12,800	12,800	12,800	12,800	12,800
Total receipts	99,200	113,600	99,200	113,600	99,200	113,600
Expenses:						
Seed, fertilizer & lime	7,804	8,128	5,320	5,320	—	—
Fuel, oil & custom hire	4,715	4,948	2,921	2,921	—	—
Depreciation, repairs, & insurance						
Buildings, storage	10,752	10,752	10,682	10,682	10,262	10,262
Power & crop machinery	8,012	8,012	7,212	7,212	—	—
Feed (including protein & urea)	4,384	5,360	19,168	22,256	59,536	62,624
Bedding	3,200	3,200	3,200	3,200	3,200	3,200
Labor	20,000	20,000	16,400	16,400	12,800	12,800
Miscellaneous	6,080	7,680	6,080	7,680	6,080	7,680
Taxes & interest						
Land/$300 A.	21,904	22,790	18,509	18,989	8,193	8,673
Land/$450 A.	27,589	28,679	22,638	23,118	8,268	8,748
Land/$600 A.	33,275	34,567	26,766	27,246	8,343	8,823
Total expenses—Land/$300 A.	86,851	90,870	89,492	94,660	100,071	105,239
—Land/$450 A.	92,536	96,759	93,621	98,789	100,146	105,314
—Land/$600 A.	98,222	102,647	97,749	102,917	100,221	105,389
Labor Income:						
Land/$300 A.	12,349	22,730	9,708	18,940	−871	8,361
Land/$450 A.	6,664	16,841	5,579	14,811	−946	8,286
Land/$600 A.	978	10,953	1,451	10,683	−1,021	8,211

called "area of economic indifference" in feeding dairy cows. This has sometimes been interpreted to mean that the substitution rates for grain and forage are such that cows can be fed any quantity of grain without great reductions in income above feed costs. The data in figure 31 illustrate several points regarding the level of grain feeding and income above

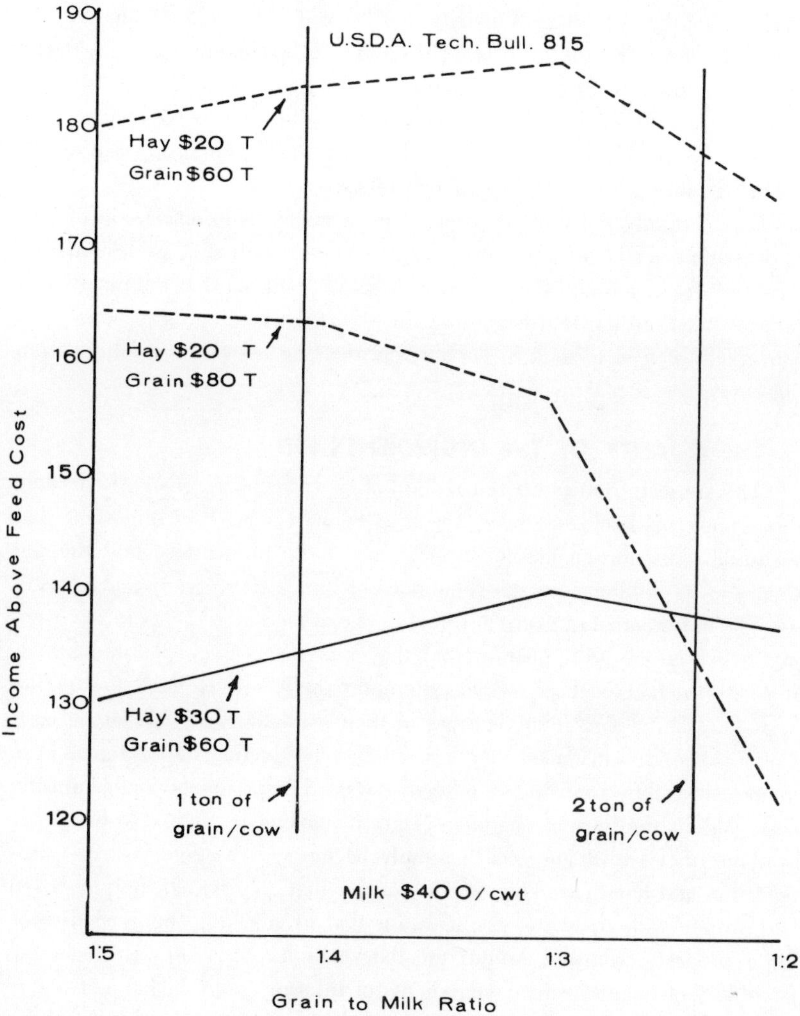

Figure 31 The Influence of Several Grain and Hay Prices and the Grain to Milk Ratio Fed on Income Above Feed Cost.

feed cost. Within this data, doubling the rate of feeding increased total grain fed by about one ton per year. Had this increase been two tons per year, the general relationship would have been reduced by about one-half. For instance, in each hay:grain price relationship, there is an area of feeding in which a change of feeding rate of about 0.5 ton grain makes little difference in income above feed cost. The area in which this occurs varies with the ingredient costs. With hay at $20 and grain at $60, this area is between a grain to milk ratio of 1:3.5 and 1:2.5. Holding hay at $20 and increasing grain to $80 moves the area to a grain to milk ratio between 1:4 and 1:5. When hay cost is raised to $30 and grain is reduced to $60, this area becomes grain to milk ratios between 1:2.5 and 1:3.5 again, only it is at a lower level of income above feed costs.

It is certainly not our intention to engage in a long discussion of how to determine feed cost–milk price relationships. Rather, it is desired that this brief discussion and these illustrations will point out that the nutritionist can never ignore the real purpose of his efforts in feeding dairy cows—the conversion of feed to milk in such a way that the net profit resulting from the enterprise is maximized.

THE QUALITY OF THE INGREDIENTS FED

This subject, from a nutritional point of view, has been constantly stressed throughout this book and need not be belabored here. It is important that we understand the economic implications of the problem so that the full weight of its influence on the economics of the dairy program can be seen.

The old practice of using low quality cheap feed ingredients in the grain ration has largely disappeared from the dairy farms of the United States. It is still true that localized farmers attempt to use locally available low quality ingredients in grain rations because of their availability at little or no cost. Under some circumstances, the loss in milk production may be good economics since they are using an otherwise wasted by-product available at little cost. Within the normal channels of grain movement, there are few feeds available that are of low quality simply because of the high cost of transportation and handling. This does not mean that all feed can be purchased and utilized with equal success: it means that by selecting the needed feeds for a properly balanced ration, the dairyman has little to fear from the possibility of obtaining feed ingredients of inferior quality. This fact is due in part to the feed laws of the nation and the normal results of competition which means that the routine selling of inferior products will soon eliminate a company from the business world.

Because of their large contribution to the total energy supply of the dairy ration and the highly variable nature of their feeding value, forages constitute the portion of the dairy ration that is most likely to influence the overall economics of the dairy feeding program. This influence may be either on the plus or negative side. So large is the influence that many dairy herds have either prospered or gone broke because of their forage programs alone. The general principles underlying the quality factor in forage have been covered in other chapters of this book. It is necessary here only to outline the results of quality differences in terms of the economics of feeding dairy cows.

The data in figure 32 illustrate the two input-output relationships which primarily influence the income above feed cost in milk production. The study

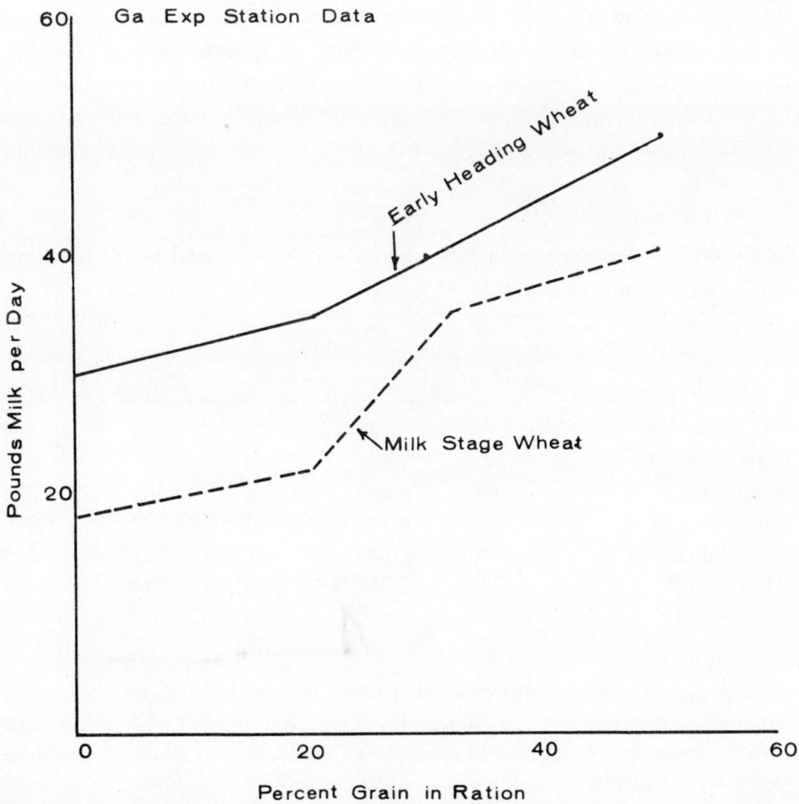

Figure 32 The Influence of Stage of Maturity at Harvest and the Percent Grain in the Ration on Total Milk Production.

was designed to compare two silages harvested from the same field and crop at two stages of maturity. The early cut wheat silage was harvested to provide a silage that was highly digestible and capable of permitting maximum intake of the total ration. The second silage (milk stage wheat) was harvested some ten days later and represents a forage that was considerably lower in digestibility and one which exerted considerable influence on intake of the total ration. The silages were fed to milking cows in combination with suitable grain rations so that 0, 20, and 50 per cent of the ration dry matter was provided by grain. The results followed closely the known principles of feed intake and animal response. Regardless of the grain fed, the superiority of the early cut silage was maintained throughout the range of feeding conditions. Adding grain produced a greater overall response with the milk stage wheat but the superiority of the early cut silage was never overcome. In essence, therefore, the lower quality silage meant that the same quantity of grain was destined to produce less milk at a greater cost at all levels of grain feeding. The data illustrate a frequently overlooked factor in the economics of feeding dairy cows: the early cut forage produced as much milk without grain additions as the late cut forage plus about 30 per cent grain. Thus in building

Table 49 The influence of hay quality on the cost and profit resulting from feeding rations capable of producing 10,000 pounds of milk.

Item	Grade of Hay				
	Excellent	Good	Average	Poor	Very poor
TDN of hay (%)	62	58	55	50	45
Gross income ($)	475	475	475	475	475
Pounds of hay[a]	6,000	5,700	5,200	4,000	3,000
Pounds of grain	1,700	2,500	3,000	4,000	7,000
Total feed costs ($)	170	173	171	187	263
Other costs ($)	235	235	235	235	235
Total costs ($)	405	408	406	422	498
Net profit ($)	70	67	69	53	−23
Value of each grade of	$25.00	$19.00	$16.00	$ 6.00	$−60.00
hay when excellent hay	29.00	23.00	20.00	11.00	−52.70
costs the amount in the	33.00	27.00	25.00	17.00	−34.00
column under excellent.	37.30	32.00	30.00	24.00	−25.30
	46.00	41.00	40.00	37.00	− 8.00
	50.30	46.00	45.00	43.00	0

[a] Pasture values equated for all hays.

a feeding program, the early cut silage provided a very high base to begin with and the grain supplementation could be used to provide for high levels of milk production instead of overcoming the basic nutritional problem resulting from the low quality silage.

The magnitude of this forage quality effect on income and hay value is further illustrated in the data in table 49. The data are a composite of many feeding trials throughout the United States and have been calculated in terms of a cow producing 10,000 pounds of milk. The data clearly illustrate the point that forage quality can alter the net profit from a positive figure to a net loss even though the cow has been fed to obtain the same level of milk production. In terms of purchasing hay, the data also illustrate the vast importance of hay quality on the dollar worth of a load of hay. Unfortunately this relationship has not been widely used in the marketing of hay with the result that purchasing hay for feeding dairy cows has frequently been a very poor management practice.

To put this phase in focus, let us consider the probable results of feeding two corn silages that were fed on Georgia dairy farms during a recent year. The silage analyses for the two farms were:

Farm No. 1		**Farm No. 2**	
Corn Silage		Corn Silage	
Dry matter	33.2%	Dry matter	43.7%
Crude protein	10.5%	Crude protein	6.5%
Crude fiber	24.4%	Crude fiber	33%
Estimated Net Energy	58T	Estimated Net Energy	38T

In comparing the two silages it is convenient to compare them in terms of providing feed for a 100 cow herd for the period required to utilize 100 tons of the poorer forage. By making this comparison, it can be pointed out that while the 100 tons of silage on Farm 2 will feed the herd for 47 days, it will require 182 tons of the better silage to do the same job since it is lower in dry matter. The feeding programs were machine calculated to provide balanced rations, but it was calculated that with the poorer silage it would not be possible even with a high level of grain feeding to produce quite as much milk as with the excellent silage.

The input-output relationships for Farm 1 were:

182 tons of silage + 32.8 tons of grain = 47 days feed.

This should produce 233,500 lbs. of milk @ $5.00/cwt =	$11,675
Total feed cost would be	3,455
Income above feed cost would be	$ 8,220

The input-output relationships for Farm 2 were:

100 tons of silage + 49.7 tons of grain	
This should produce 219,490 lbs. of milk @ $5.00/cwt =	$10,975
Total feed cost would be	3,992
Income above feed cost would be	$ 6,983

This comparison is no different from the thousands of such comparisons which have come from farm records throughout the United States. It is somewhat surprising that it is still so little heeded on many dairy farms. The prevalent practice of measuring crop yields and success in terms of the amount of feed harvested from a field or the small acreage required to fill a silo is ample evidence that we need to constantly keep in mind the all important factor in feeding dairy cattle—INCOME ABOVE FEED COST.

PROPER TIMING OF FEED USE

No matter how good the feed or how low the cost, feed fed at the wrong time in the lactation may well wipe out the savings in feed cost from otherwise good management. This brings into play both the normal lactation curve of dairy cows and the tendency for improper feeding at one stage to inhibit a good response at another stage of lactation. This portion of dairy feeding is probably the most frequently encountered failure observed in the field. When pasture was used as the primary source of summer feed, literally thousands of pounds of milk were lost each year from a failure to understand this principle. Farmers, by nature, are hopeful people. A frequently encountered situation went like this. It was a hot August and the pasture was abundant but poor and the herd was calving to begin the base-milk period. The dairyman, not understanding the detrimental influence of low quality forage on the entire ration, was attempting to feed his fresh cows extra grain to initiate a high level of milk production. Sometimes extra feed in the form of hay or silage was being offered but the cows were consuming quantities of low quality pasture which was limiting both the intake of good forage and the feeding value of the entire ration. The result: less than expected milk production. When the cows were removed from the pasture and fed good silage, higher levels of milk production were produced with even lower levels of grain feeding. Over a total lactation, it is not unusual to find herds with 10,000 pound herd averages feeding 5000 pounds of grain living side-by-side with other herds with 14,000 pound herd averages on 4000 pounds of grain per cow. Many of the causes for this difference are illustrated in the series of charts which follow. The charts are greatly smoothed-out

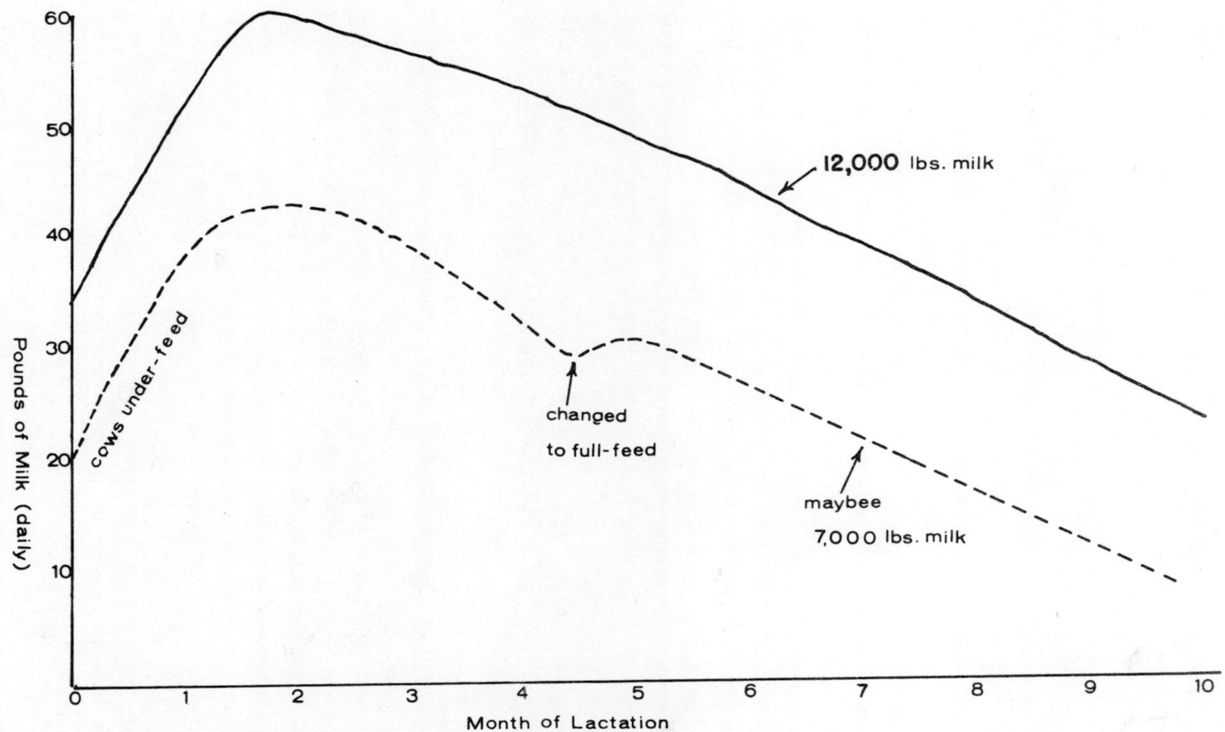

Figure 33 The Effect of Under-feeding During the Early Part of the Lactation and Changing to Full Feeding in Mid-lactation.

versions of normal lactation curves since the actual curves would include many temporary "ups" and "downs" due to normal fluctuations in milk of individual cows.

The illustration in figure 33 shows the effects of under-feeding cows during the early part of the lactation and then attempting to correct the problem by changing to normal feeding later in lactation. This is rather typical of what happens when a cow freshens without enough body condition to permit her to reach a normal peak of milk production by using her body stores. The result is a lower than average initial level of production and a tendency for an early decline in lactation. Halfway through the lactation the cow is changed to a high plane of nutrition and responds with an increase in milk production. The response is small since the cows respond in proportion to their current level of milk and the rate of decline is now greatly influenced by the fact that the cow is pregnant and her body is equally concerned with preparing for the subsequent lactation. The result is a loss of 4,000 or 5,000 pounds of milk during the lactation.

Another case is illustrated in figure 34. This is frequently seen in herds where grain is fed on the basis of milk production with the result that production controls grain feeding instead of grain feeding leading production. The cows that were fed high quality silage produced 12,000 pounds of milk even with a set grain to milk feeding ratio since the high quality forage permitted many improvements by allowing greater levels of intake. As was pointed out in figure 32, when poor silage was fed a great portion of the grain fed went to compensating for the lower quantity of nutrients supplied by the forage. The cows were obviously in good body condition since they peaked normally, but their rate of decline was exaggerated and the total lactation was reduced by some 3,000 pounds of milk. Under such conditions, the cow just never had a chance.

The lactation curve shown in figure 35 was once the "American" way of producing milk and is still encountered in some grade "B" markets and in several areas of the world. This was and is the program that makes maximum use of forage by going for maximum use of forage per acre. The cows are freshened on lush spring pasture and produce very well for two or three months. During this period excess pasture and fields set aside for hay are harvested for winter feed. Since it is necessary to produce a maximum of hay for winter feeding, the forages are permitted to reach an advanced stage of maturity to insure maximum yields per acre. When pasture quality declines in late summer, the milk flow declines rapidly. With the onset of winter, the cows move to the low quality hay. Many farmers begin a high

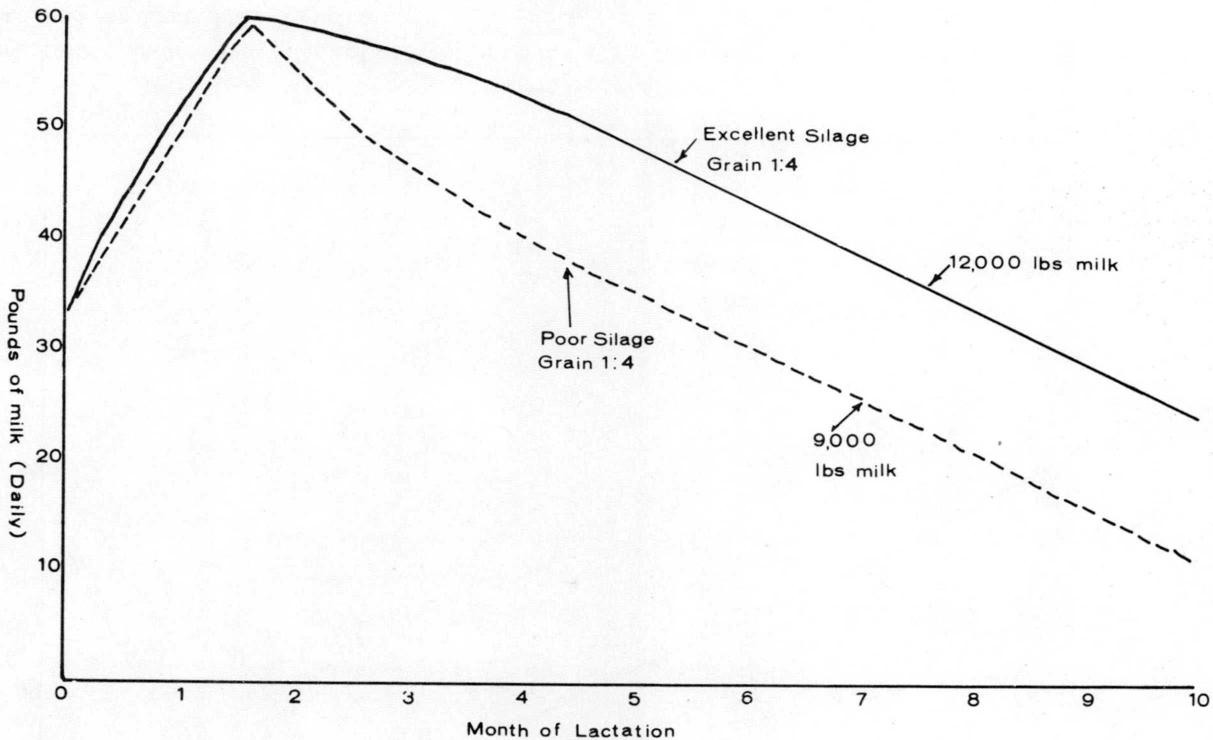

Figure 34 The Influence of Forage Quality on the Lactation Curve When Grain is Fed on the Basis of Milk Produced.

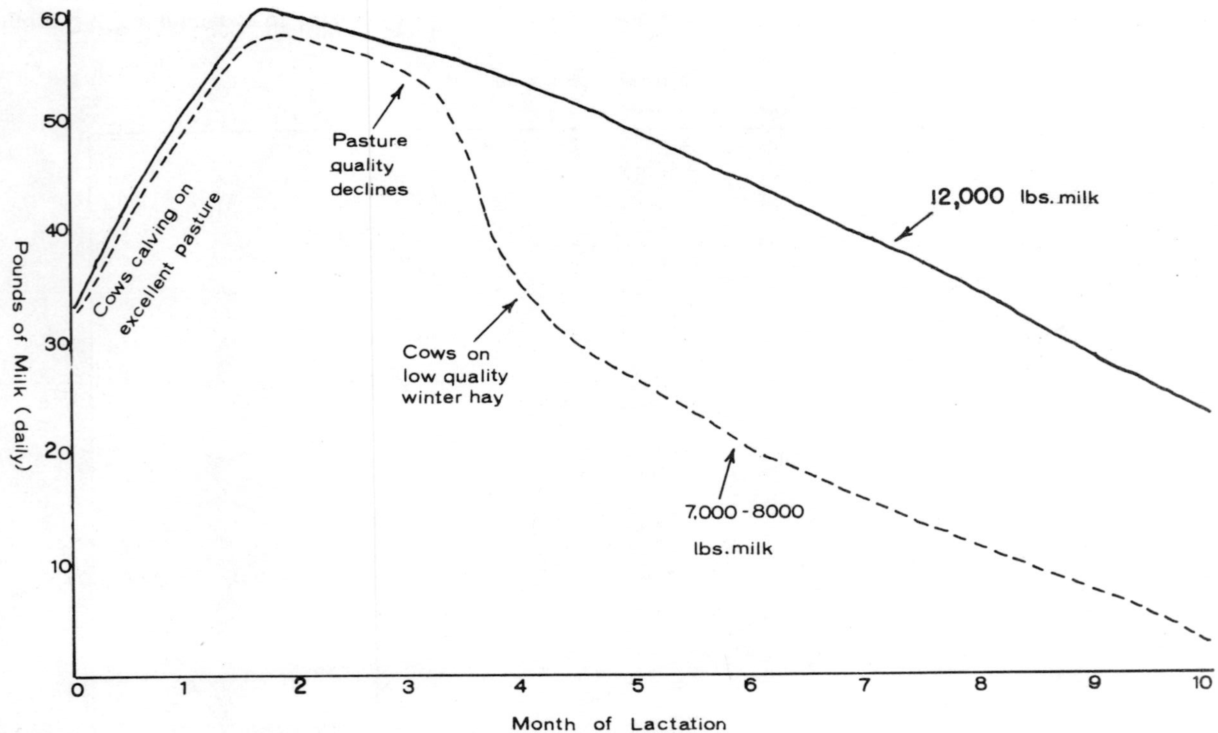

Figure 35 A Lactation Curve Typical of That Observed When Maximum Utilization of Forage Per Acre is the Object Rather than Maximum Milk Production.

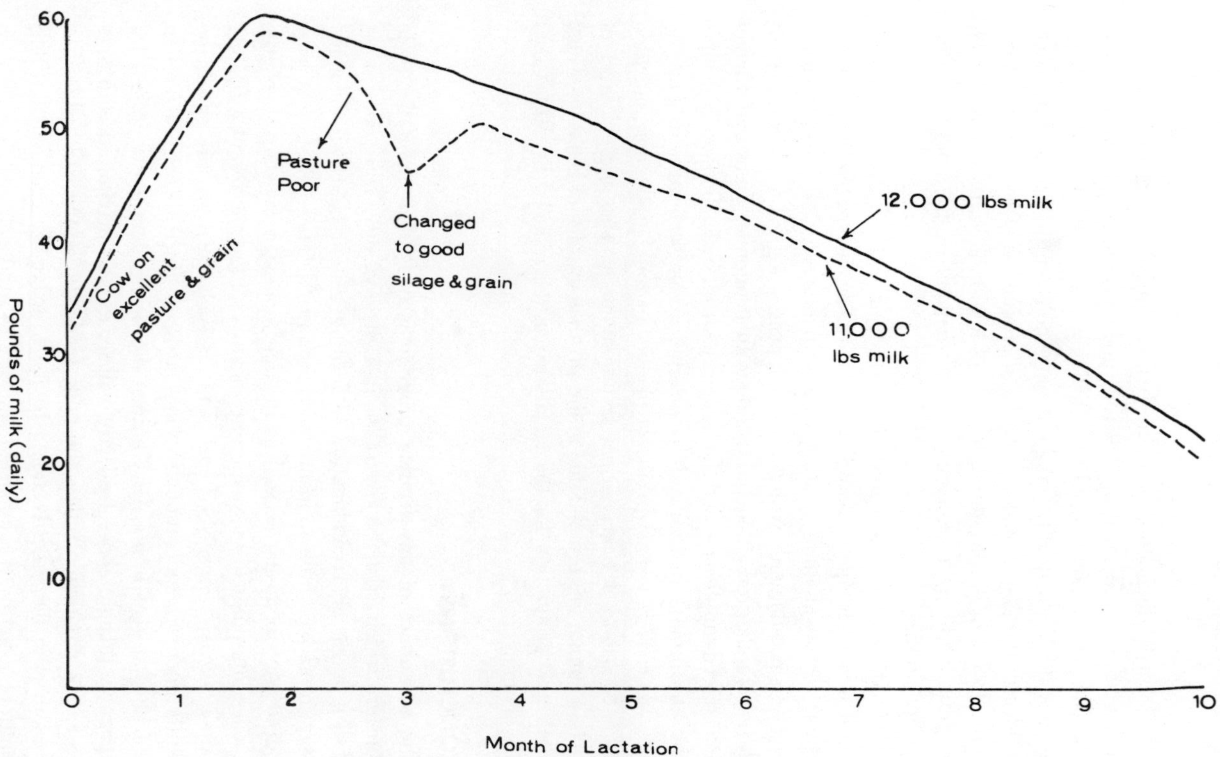

Figure 36 The Long-time Influence of Short-time Under-feeding Due to An Emergency Such as Drought.

level of grain feeding along with the hay but this has little real influence on the lactation curve since the level of production has declined to a point that the cow will respond very little, particularly since the remainder of the ration consists of low quality forage. If only the pasture portion of the curve is considered, the system may show a large production of cheap milk. When the entire lactation is considered, the loss of total milk and the considerable cost of grain fed to low producing cows will frequently show a net loss rather than a net profit.

The final illustration in figure 36 illustrates a factor that can sometimes be useful under emergency conditions. This chart illustrates the influence of a short period of under-feeding within an otherwise well fed lactation. This curve is possible only with cows which possess a great tendency towards high levels of milk production and is of doubtful economics. In case of drought or reduced feed supplies beyond the farmers control, such a temporary decline may be an absolute necessity. It is possible under such circumstances to permit a short (2-3 weeks) period of decline in milk production and then realize a return to a near normal lactation curve. The original level of production cannot be regained due to the advance in stage of lactation. Thus, the total milk production for the lactation may be reduced by 500–1,000 pounds of milk.

CONCLUSIONS

The economics of any feeding program is dependent upon the price relationships which exist at a given time in a given locality. Thus, one cannot establish a single or a combination of feeding systems and call them the best way to produce milk to insure maximum profit. There must be at least a hundred ways of losing as well as making money in the dairy business. From a feeding point of view, the one measure of the success of a feeding program is *income above feed cost,* and, as such, this measure should be constantly under consideration by dairy farmers. Since so large a fraction of the feed that goes into a dairy animal during her lifetime is utilized for maintenance, income above feed cost will usually be maximized with high levels of milk production unless the milk : feed ratio is unusually wide. For this reason, most dairy herds should be fed to maximize production within the limits imposed by the inherited capacity to produce.

The dairy enterprise should be utilized primarily to convert home-grown roughage into milk. In making this conversion, constant emphasis should be placed on using only high quality feed fed at the proper time to produce maximum response. If unusual economic conditions are encountered which indicate feeding for something less than maximum production, the long time consequences should be considered in making decisions since once a lactation curve is permitted to decline, it cannot be renewed during the current lactation.

Finally, the dominant role of the forage portion of the ration should be recognized. Of all the potential errors in feeding dairy cows, the error resulting from utilizing grain to compensate for the low feeding value of forage must rate as potentially the most costly.

References

1. Hoglund, C. R. Economics of Growing Vs. Buying Feed for Dairy Herds of 40 to 240 Cows. Agricultural Economics Report No. 83, Michigan State University. 1967.
2. McCullough, M. E. and L. R. Sisk. Influence of Stage of Maturity at Harvest and Level of Grain Feeding on Intake of Wheat Silage. J. Dairy Science 50:705. 1967.
3. Moore, L. A. The Place of Liberal Concentrate Feeding to Dairy Cattle. U.S.D.A., ARS 44–133. 1964.

Chapter 8

Feed Programming of Optimum Rations

So far, we have been concerned with the principles of feeding dairy cows and the development of the concept of optimum feeding. All of the feeding programs and the principles of feed programming are useless until they are transformed into a ration which is fed to a cow to produce milk. Thus we have reached the point where we must consider the mechanics of transforming principles into practice—in this case, optimum rations.

This demand for a new method of feeding is the natural result of several developments in the livestock field. The conditions which have developed are similar to the ones predicted some years ago by Dr. Hutton in his series of articles on "Feed Programming." He took a look into the future and commented: "Let us take the problem of feeding dairy cattle, for instance. Several states, including Pennsylvania, are planning, or have underway, a forage testing program. When such a program becomes of general use, which seems to be a matter of relatively short time, farmers will know more precisely what they are feeding in the way of forage. The old crude approximations based upon species, color, leafiness (and even the more recently developed crude approximation based upon date of cutting) will give way to a more objective measure of feeding value. As long as the farmer had only the vaguest notion of his forage analysis, he was content to accept rough rules-of-thumb concerning the level and composition of the grain component of the ration. If he was not content, at least there was little that he could do about it. Prudent husbandry dictated that he cover risk of nutritional deficiency by a wide and costly safety margin in the grain mix specifications. Moreover, in the presence of a wide area of ignorance, there was little point in being precise in the grain mix specifications. He generally was willing to accept one of the "standard model" feed mixes provided by the feed milling industry.

This attitude should change with the coming of forage analysis. We expect the farmer to look upon the feeding problem as one of meeting the "specifications" set by the need of the animals he is feeding. His home produced forages (and other feeds) will be one set of ingredients that can be used to meet these specifications. He will desire of his grain supplement the least-cost means of satisfying the residual specifications."

In a sense, the above predictions of Dr. Hutton have now come true on thousands of dairy farms in several areas of the United States. It is anticipated that what is now a trend will become the major method of feeding dairy animals throughout the United States and in many other countries.

THE CONCEPT OF AREA OF NUTRITION

Throughout the previous chapters of this book, much has been said about the many variables in feeding dairy cows and the wide range of animal response to even the same ration. The basic concept of optimum rations includes these variations and provides for their partial solution by free-choice feeding of the rations to permit the cows to exhibit their full individuality. In feed programming, it is necessary to define nutrient requirements, feed composition, and other items in precise mathematical terms. On the surface this may seem to be a contradiction of all we know about animal and feed variation. If we treat this precise calculation as though it were a precise ration that had fixed characteristics such as intake, digestibility, and feeding value, then we do ignore all of the variations. On the other hand, if we treat this precise method of calculating rations as a means of producing a ration that is within a workable range of the needs of the animals with their variations considered, we will be closer to the facts which will prevail in the feedlot. To do this, the author has developed the habit of thinking of animal requirements in terms of areas of nutrition rather than as fixed and definable requirements. Thus, for each type of animal performance, there are three nutritional areas: under-nutrition, optimum nutrition and over-nutrition. It is not necessary to draw fine lines between the areas but it is desirable that the rations provided fall somewhere in the high optimum to low over-nutrition area when high levels of production are the object and in the low-optimum to high under-nutrition area when only maintenance is desired. In practice, the range in energy content as represented by digestibility of the ration will be 4 to 6 percentage points. For instance, if the true desired digestibility of the ration is 65 per cent, rations that will probably provide similar animal response can vary from say 62 per cent to 68 per cent in digestibility. Other variables such as intake differences and variations in true animal requirements will exert influences greater than digestibility. This concept is important since it must be obvious that normal variations in the chemical composition of feed ingredients, the digestibility of similar lots of the same ingredient, and the variations in intake and utilization of similar diets by similar animals are such that precise calculations cannot be made of either ration composition, ration intake or ration digestibility. When this is kept firmly in mind, the manipulator of feeds and formulas in feed programming will not make the error of believing that his 70 Therm ration actually always has 70 Therms or that his cows which have a calculated protein requirement of 14 per cent actually always have such a requirement. He will recognize that in feed programming one can only calculate rations that are within the "area" of the requirements of his animals and that final adjustments must be made in the feedlot to provide the level of performance that he desires.

CALCULATING OPTIMUM RATIONS

Before demonstrating the calculation of an optimum ration, it would be useful to introduce and describe a most useful technique in ration calculations. The technique is frequently referred to as the Pearson Square, although the author has been unable to determine its origin. In practice, it has been used for years in dairy manufacturing to determine the relative proportions of ingredients in dairy mixes such as in ice cream manufacture. A square and its use is illustrated in figure 37. In this case the problem is to combine corn with 8 per cent crude protein and soybean meal with 50 per cent crude protein to provide a mix with 20 per cent crude protein. This is done by

PROBLEM: Combine corn with 8% crude protein and soybean meal with 50% crude protein to provide a mix with 20% crude protein.

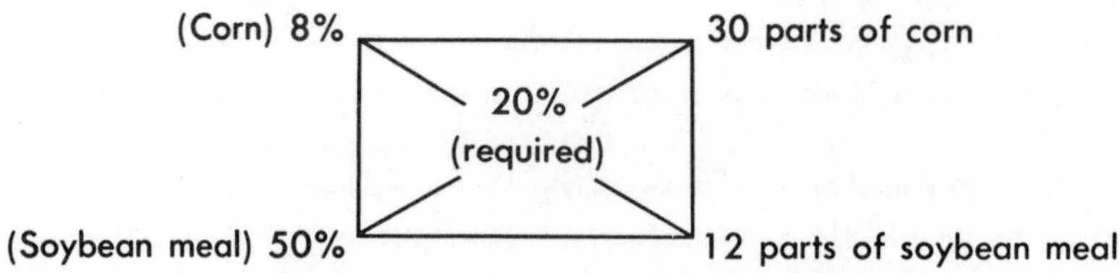

42 parts of ration

$$\frac{30}{42} \times 100 = 71.4\% \text{ corn}$$

$$100 - 71.4\% = 28.6\% \text{ soybean meal}$$

Figure 37 An Example of the Construction and Use of the Pearson Square in Solving Feed Formulation Problems.

writing the required protein content 20 per cent in the center of the square and the protein content of the two ingredients at the left side corners of the square. The proportion of each ingredient is determined by recording the numerical difference between the values on the left hand of the square and the value in the center. Signs and other mathematical considerations are ignored. In this case the difference between 50 per cent and 20 per cent is 30 and the difference between 8 per cent and 20 per cent is 12. The values are written diagonally and indicate that the mix should contain 30 parts of corn and 12 parts of soybean meal. The sum of the two proportions ($30 + 12 = 42$) can be used to convert the proportions to percentages for easier calculation. By dividing the parts of corn (30) by the sum (42) and multiplying by 100, it is learned that the mix should contain 71.4 per cent corn. The percentage of soybean meal required is determined by subtracting the required percentage of corn (71.4) from 100—the percentage being 28.6 per cent. Thus, one hundred pounds of the mixture could be obtained by mixing 71.4 pounds of corn with 28.6 pounds of soybean meal.

Now we can proceed to formulate an optimum ration of milking dairy cows from the knowledge available in the general description of optimum rations in Chapter 8 and the results of forage testing. We will use as an example a silage sample from a dairy farm in Georgia that was submitted to us for this purpose. We have selected a rather poor sample since it will enable us to demonstrate several principles other than the mere mechanics of calculating the optimum ration.

FORAGE ANALYSIS

Type of forage: sorghum silage

Moisture Content: 64 %

Dry matter content: 36 %

Forage Analysis (dry matter basis)

1. Crude Protein: 9.9. %
2. Digestible Protein: 4.7 %
3. Crude Fiber: 28.5 %
4. TDN: 56.8 %
5. ENE: 44.6 T/100 lb.

The problem is to utilize the above forage analysis to formulate a total ration containing 14 per cent crude protein, 18 per cent crude fiber, 60 T Net Energy, and to describe the grain ration and recommend the proportions of grain and silage to be mixed for an optimum ration.

Step 1 The first step is to determine the percentage of grain and silage dry matter that will be required to provide the 60 T ration desired. Since the silage is relative low quality, it can be seen that a ration with at least 70 T Net Energy should be used. Having made this decision, the square method can be employed to determine the proportions of grain and silage needed.

Grain	70		15	(60% grain)
		60		
Silage	45		10	(40% silage)
			25	

Step 2 The second step is to determine the per cent protein needed in the grain ration to insure a total ration with 14 per cent protein. This can be accomplished by first recording the known information and then solving for the unknown percentage. Thus:

40% silage × 10% protein contributes 4.0% protein
60% grain × ____ protein contributes 10.0% protein
Protein percentage required 14.0

By dividing the percentage of grain in the ration into the percent of protein that must be contributed by the grain (10 divided by 60) it is found that the grain ration should contain 16.6 per cent protein.

Step 3 The third step is to determine the percentage of crude fiber that must be included in the grain ration to provide 18 per cent crude fiber in the total ration. This is acomplished in the same manner as for the protein requirement.

40% silage × 28% crude fiber contributes 11.2% fiber
60% grain × ____ crude fiber contributes 6.8% fiber
Crude fiber percentage required 18.0

By dividing the percentage of grain in the ration into the percentage of fiber it must provide we find that the grain ration should contain 11.3 per cent crude fiber.

Step 4 The grain ration can now be formulated in the usual way to meet the requirements which are 16.6 per cent crude protein, 11.3 per cent crude fiber, and 70 T Net Energy per 100 pounds of ration. One combination of ingredients that meets the requirements and the limitations imposed on the ingredients used is:

Ingredient	Pounds	Crude protein	Crude fiber	Net Energy
		lb.	lb.	Therms
Wheat Bran	100	16	10	60
Beet Pulp	350	32	66	238
Citrus Pulp	350	23	46	262
Brewers Dried Grains	350	91	53	220
Corn	550	51	11	440
Cottonseed Meal (41%)	300	123	33	204
Total	2000	336	219	1424
Percentage		16.8	11	70

If the silage is calculated on an as fed basis, the recommended optimum ration would be formed by mixing 2222 pounds of silage and 1200 pounds of the above grain ration.

This method has resulted in the formulation of a grain ration that may appear to be composed of an unusual combination of ingredients but it is typical of the type of ration that is required to maintain normal milk production and composition when it is necessary to feed so large a proportion of grain to silage due to the low quality of the silage. It also illustrates the fact that a high energy ration can be formulated from ingredients that have considerable bulk. This is important since many people confuse "high energy" rations with "high density" rations containing primarily corn and oil seed meals.

The grain mixture formulated is obviously only one of literally hundreds that could be used. The mixture meets only three of the many requirements that could be listed for formulation but it does about exhaust the limitations that can be handled on desk calculators. The other obvious question is the cost of the ration. In light of the ever-present need for economy in feeding cattle, the question is very important. For these reasons, there is much current interest in using computers to formulate rations, thus increasing the ingredients that can be considered, the number of restrictions that can be placed on the ration and to insure economy by calculating "least-cost" rations.

LEAST-COST OPTIMUM RATIONS

The idea of programming feeding operations to insure least cost rations is very appealing to livestock farmers. At the same time, most farmers understand that one cannot really provide a "least-cost" ration but rather a least-cost ration that is balanced in terms of nutrition for their animals and also a ration that animals will eat. To do this successfully requires many more calculations than most people can do or are willing to do using a pencil or even a desk calculator. For this reason, in recent years much has been heard about computer rations and the use of high-speed electronic computers to provide feeding programs. There is still something magic about the word "computer" and many people, including nutritionists, appear to accept the idea that just because a ration was calculated on a computer that it is good and should be used without question. At the beginning, one should understand that *rations or feeding programs formulated on a computer are no better than the information provided the computer for the formulation.*

The method employed to formulate least-cost rations with computers is known as *Linear Programming.* Technically, the definition of linear programming is as follows: "A technique for obtaining a unique value-weighted solution to a set of simultaneous linear equations in which the number of unknowns may exceed the number of equations and in which no variate has a negative value." We can demonstrate the basic idea in linear programming by a rather simple example.

A farmer needs a ration containing 14 per cent protein and 65 per cent TDN. To make this ration he has two feeds—corn with 8 per cent protein and 80 per cent TDN and alfalfa hay with 16 per cent protein and 55 per cent TDN. To solve this problem with simultaneous equations one would write the problem as follows:

$$.08a + .16b = .14$$
$$.80a + .55b = .65$$

The top equation indicates that some proportion of corn (a) plus some proportion of alfalfa hay (b) can be combined to secure the 14 per cent protein needed. The percentages have been converted to portions of a pound which is the usual unit of measure in linear programming. Likewise, the second equation indicates the same thing for the TDN requirement. Using the principles of algebra, the two equations can be solved simultaneously to show that a = .325 and b = .714. Converting the two values to percentages, the solution to the problem is that a mixture of 31.2 per cent corn and 68.8 per cent alfalfa hay will provide a ration with 14 per cent protein and 65 per cent TDN. So far, this is a solution that could have been just as well solved by the Pearson square method.

Now, lets add another feedstuff. In addition to the corn and alfalfa hay, lets add corn distillers dried grains with 27 per cent protein and 80 per cent TDN. Now the equations look like this:

$$.08a + .16b + .27c = .14$$
$$.80a + .55b + .80c = .65$$

Now, we are in trouble for there are obviously many combinations of the three feedstuffs that can be combined to provide the ration with 14 per cent protein and the 65 per cent TDN. Hutton pointed out that in a relatively simple mix problem with six quality specifications and 10 ingredients, there could be as many as 34 trillion feasible solutions. There is a way to reduce the number of solutions to only one—that is to add prices for the ingredients and limit the mix to a least-cost formulation. If we give a value to the ingredients of $.02 per pound for corn, $.01 per pound for hay and $.03 per pound for the corn distillers grains, the equations now become:

$$.08a + .16b + .27c = .14$$
$$.80a + .55b + .80c = .65$$
$$2a + 1b + 3c = \text{minimum cost}$$

The solution to this set of simultaneous equations involves the use of matrix algebra and we enter the world of computers. When we do, we should understand enough about what we are doing to make intelligent use of the techniques although there is little need for the nutritionist to know the mathematical details or the methods used to program machines and the details of machine operation.

The least-cost linear programming matrix can be divided into three major areas for discussion: ration requirements and restrictions, ingredient analysis, and objective function of ingredients. The interaction of these three can best be illustrated by an example. We programmed least-cost formulations for optimum rations for three corn silages which will be utilized to illustrate several of the problems in linear programming of optimum rations.

Ration requirements and restrictions The requirements and restrictions for the rations are shown in table 50. It can be seen that the values can be set as maximum values, minimum values or as exact values. Thus, in the first run, the ration was set at exactly 2000 pounds of feed with a minimum of 260 pounds of protein and a maximum of 280 pounds of protein;

Table 50 Ration restrictions for runs 1 and 2.

Item	Restrictions per 2000 pounds
First Run	
Cost	Minimize
Total pounds	Exactly 2000 pounds
Crude Protein	Not less than 260 pounds
Crude Protein	Not more than 280 pounds
Crude Fiber	Not less than 360 pounds
Crude Fiber	Not more than 400 pounds
Metabolizable energy	Not less than 2000 Mcal
Metabolizable energy	Not more than 2650 Mcal
Calcium	Not less than 18 pounds
Calcium	Not more than 20 pounds
Phosphorus	Not less than 12 pounds
Phosphorus	Not more than 14 pounds
Second Run	
Total pounds	At least 2000 pounds
Metabolizable energy	At least 2550 Mcal
Metabolizable energy	Not more than 2650
All other restrictions same as in first run.	

crude fiber, minimum of 360 and maximum of 400 pounds; metabolizable energy, minimum of 2000 Mcal and maximum of 2600 Mcal; calcium minimum of 18 and maximum of 20 pounds; and, phosphorus at a minimum of 12 and maximum of 14 pounds.

In the second run, the restrictions on metabolizable energy were narrowed to a minimum of 2550 Mcal and a maximum of 2650 Mcal. With this narrow restriction, the problem was not feasible of solution with an exact requirement of 2000 pounds of feed so this restriction was changed to a minimum of 2000 pounds of feed.

Ingredient analysis In table 51 are listed the ingredients, their analysis, their cost and their restrictions. It is, of course, necessary that all values be listed in terms of a common unit—in this case, on a per pound basis, The importance of this part of programming cannot be over-emphasized. Initially, it must be remembered that this list of ingredients must be such that the machine can formulate a feeding program within the limitations

Table 51 Feed ingredients: Their composition, cost and restrictions for optimum rations.

Ingredient	Crude protein	Crude fiber	Calcium	Phosphorus	Metabolizable energy	Cost	Restrictions
				per pound			
1. Alfalfa Meal	.174	.248	.0132	.0032	.95	.027	Not over 100 lb.
2. Beet Pulp	.091	.190	.0068	.0010	1.13	.031	Not over 300 lb.
3. Brewer's Dried Grains	.259	.150	.0027	.0050	1.08	.031	Not over 400 lb.
4. Citrus Pulp	.066	.130	.0196	.0012	1.37	.022	Not over 100 lb.
5. Corn	.093	.020	.0002	.0029	1.45	.019	None.
6. Corn Distillers Grains	.270	.093	.0010	.0037	1.40	.035	None.
7. Corn Gluten Feed	.253	.080	.0046	.0077	1.36	.029	Not over 500 lb.
8. Cottonseed Meal (41%)	.416	.110	.0015	.0110	1.15	.041	None.
9. Oats	.118	.110	.0010	.0035	1.09	.035	Not over 200 lb.
10. Soybean Meal (50%)	.500	.020	.0033	.0062	1.09	.044	None.
11. Alfalfa Haylage	.178	.302	.0137	.0045	0.95	.015	Not over 500 lb.
12. Dicalcium Phosphate	—	—	.2400	.1800	—	.050	None.
13. Defluorinated Phosphate	—	—	.3200	.1800	—	.050	None.
14. Bone Meal	—	—	.2900	.1300	—	.050	None.
15. Corn Silage No. 1	.067	.330	.0049	.0009	.901	.010	Not over 1200 lb.
16. Corn Silage No. 2	.101	.219	.0033	.0023	1.18	.010	Not over 1200 lb.
17. Corn Silage No. 3	.092	.259	.0033	.0023	1.08	.010	Not over 1200 lb.

imposed. For instance, the ingredient list must contain feedstuffs that are as high or higher in energy than the minimum set in the restrictions. What is true for energy is true for all other requirements. It must also be remembered that the more requirements that are included, the greater the range in nutrient composition which must be available if a ration is to be calculated. An alternative to the use of many ingredients is to limit the requirements to one or two important ones, have the machine calculate the quantities of minor items available (as an example—the quantity of calcium and phosphorus in the ration), and simply add enough of that nutrient to obtain the quantity needed. In this case, the three corn silages were listed as ingredients 15, 16, and 17. In practice, these would be programmed into the machine as an added ingredient for each feeding program. They are listed here to illustrate the fact that in this type of programming, the silage component is utilized as just another feed ingredient.

Objective function of ingredients The objective function is the factor that correlates the two sides of the problem (it combines the available feeds to provide a ration that meets the requirements of the animals). It must be clear by now that this final solution is no better than the requirements set and the validity of the composition of the feeds available. Of importance, is the fact that the ration will always be least-cost. This is not as good as it may sound on first reading. Most feeders and all experienced nutritionists know that the least-cost ration is not the one that always provides the most economical animal performance or the best rate of feed conversion. The best ration may be near the least-cost one but it may not be possible to reduce all nutritional knowledge into mathematical restrictions which can be placed in the program. For instance, ingredients tend to interact to some degree (corn distillers grains interact to a degree with corn silage to provide a better ration than can be described with numerical values), efficiency of ration utilization depends to a degree on level of intake which may be influenced by the ration although it cannot be programmed in light of our present state of knowledge, and, finally, not all relationships between ingredients are linear. These limitations to the use of linear programming should not be seen as an argument against its use: they should be seen as an additional requirement. Sound nutritional judgment must be exercised in using the programmed rations and they should be altered to fit that judgment.

The rations in tables 52, 53, 54, and 55 illustrate the results of the several feed programs calculated by the computer when variations were made in the ration requirements and restrictions.

The rations in table 52 and in the first column of table 55 illustrate

Table 52 Optimum rations as influenced by ration restrictions.

Item	Programmed with requirements for:			
	Energy, protein, fiber, minerals	Energy, protein, fiber	Energy, protein	Energy, protein, fiber minerals (no cost)
RUN 1—SILAGE 1				
Composition of Optimum Ration:				
Corn Gluten Feed (pounds)	—	—	377.73	—
Corn (pounds)	624.25	671.02	—	621.74
Soybean Meal, 50% (pounds)	130.91	122.52	.54	131.36
Alfalfa Haylage (pounds)	500.00	500.00	500.00	500.00
Dicalcium Phosphate (pounds)	30.22	—	—	37.99
Defluorinated Phosphate (pounds)	5.83	—	—	—
Silage dry matter (pounds)	708.77	706.42	1121.68	708.93
Cost per ton of ration:	$ 34.01	$ 32.70	$ 29.69	$ 32.18
Composition of Optimum Ration:				
Crude Protein (%)	13	13	13	13
Crude Fiber (%)	20	20	27	20
Energy (Mcal/kg)	2.4	2.4	2.2	2.4
Calcium (%)	1.0	.54	.7	1.0
Phosphorus (%)	.6	.28	.3	.61

the programs obtained with silage 1, which was a silage with a low feeding value. Using the lower restrictions on energy, the machine found its least-cost ration by making maximum use of alfalfa haylage to produce a ration with maximum fiber and minimum protein content. By raising the energy requirements (table 55), the machine solved the problem by choosing citrus pulp and distillers grains to replace alfalfa haylage, but now provided a ration with maximum protein and minimum fiber. This illustrates an important point. When one restriction (in this case—energy) is changed, this may result in vast changes in other components of the ration since the controlling factor is always a search for least cost.

In the second column of table 52, the requirements for calcium and phosphorus were removed. In this and all other rations, this made little real difference in the rations. Since the minerals were being added from ingredients without energy, protein or fiber contents, this small change would be expected.

Removing the fiber restrictions (column 3, table 52) resulted in drastic changes in the optimum ration. The machine further dropped the energy content of the ration and proposed a ration with 27 per cent crude fiber and a total of better than 80 per cent of the total ration from corn silage and alfalfa haylage. This ration would obviously not work since its total intake would be below the requirements for high levels of milk production. Here we can illustrate another point in writing restrictions. When two or more forages are involved, a restriction must be included on the total amount of forage in the ration should they be used in combination.

When the mineral requirements were returned to the program but the mineral sources were entered at no cost (column 4, table 52) and the fiber restriction was added, the ration was little different from the first, again showing the minor influence of low-cost ingredients used at low levels.

The rations for silage 2 (table 53 and column 2 of table 55) illustrate the results of similar manipulations of the requirements for the rations in the case of a high quality corn silage as the basic least-cost ingredient. The

Table 53 Optimum rations as influenced by ration restrictions.

	Programmed with requirements for:			
Item	Energy, protein, fiber, minerals	Energy, protein, fiber	Energy, protein	Energy, protein, fiber, minerals (no cost)
RUN 1—SILAGE 2				
Composition of Optimum Ration:				
Corn (pounds)	258.37	301.74	246.19	249.24
Soybean Meal (50%) (pounds)	75.68	68.47	53.81	77.20
Alfalfa Haylage (pounds)	432.18	429.78	500.00	432.68
Dicalcium Phosphate (pounds)	12.31	—	—	40.86
Defluorinated Phosphate (pounds)	21.43	—	—	—
Silage dry matter (pounds)	1200.00	1200.00	1200.00	1200.00
Cost per ton of ration:	$ 28.41	$ 27.19	$ 26.54	$ 26.62
Composition of Optimum Ration:				
Crude Protein (%)	13	13	13	13
Crude Fiber (%)	20	20	21	20
Energy (Mcal/kg)	2.5	2.6	2.5	2.5
Calcium (%)	1.0	.5	.55	1.0
Phosphorus (%)	.6	.3	.3	.66

Table 54 Optimum rations as influenced by ration restrictions.

Item	Programmed with requirements for: Energy, protein, fiber minerals	Energy, protein, fiber	Energy protein	Energy, protein, fiber, minerals (no cost)
RUN 1—SILAGE 3				
Composition of Optimum Ration:				
Corn (pounds)	366.01	409.87	219.65	348.46
Soybean Meal (50%) (pounds)	137.85	130.55	80.34	140.77
Alfalfa Haylage (pounds)	261.99	259.57	500.00	262.96
Defluorinated Phosphate (pounds)	34.13	—	—	—
Dicalcium Phosphate (pounds)	—	—	—	39.18
Steamed Bone Meal (pounds)	—	—	—	8.61
Silage dry matter (pounds)	1200.00	1200.00	1200.00	1200.00
Cost per ton of ration:	$ 30.65	$ 29.42	$ 27.20	$ 28.75
Composition of Optimum Ration:				
Crude Protein (%)	13	13	13	13
Crude Fiber (%)	20	20	23	20
Energy (Mcal/kg)	2.4	2.5	2.5	2.4
Calcium (%)	.95	.4	.55	1.0
Phosphorus (%)	.60	.3	.20	.7

major item is again the probable detrimental influence on intake resulting from a failure to program a requirement for a maximum forage allowance. When the energy requirement was raised (table 55), the alfalfa haylage disappeared and the corn was replaced by citrus pulp and distillers grains. Actually, with a very minor amount of alteration, any of the 5 optimum rations could probably be made into very useful rations. Thus, the higher the quality of the feedstuffs employed—particularly the forages—the easier it is to write realistic requirements and restrictions for least-cost programming.

The rations for silage 3 (table 54 and column 3, table 55) are intermediate between those of silages 1 and 2. They would appear to need no special comment.

All three rations in table 55 suffer from the lack of an exact limitation on total pounds of feed. The machine successfully placed 200 pounds of protein, 360 pounds of crude fiber, 20 pounds of calcium and 12 pounds of phosphorus in each ration. However, the percentages of nutrients are not the same since the silage 1 ration contains 2130 pounds, silage 2, 2067 pounds, and silage 3, 2021 pounds. In this case, the differences would not be expected to result in vast disappointments in the feed trough since all the rations are high in total feeding value. The data do indicate the need for always providing for an exact requirement for total ration.

The above examples are not presented to discredit linear programming of least-cost optimum rations. They are presented to illustrate the necessity for putting maximum effort into the preparation of both the ration restrictions and requirements, and the list of usable feedstuffs so the computer can formulate realistic and usable feeding programs. Failure at any point in the process will result in poor animal performance which is due to the information provided the computer, and not to the fact that the animals are being fed "computer" rations.

CONCLUSION

It is obviously a long step from the long-time recommendations for careful calculation of rations for individual cows by considering their body weight, milk production, and fat test to the idea presented here of mixing one ration to be self-fed to all cows in a herd regardless of the variations in their nutrient

Table 55 Run 2. The rations and their composition as influenced by silage quality.

Item	SILAGE 1	SILAGE 2	SILAGE 3
Ration Ingredients:			
Citrus Pulp (pounds)	283.42	296.54	348.98
Corn (pounds)	391.97	4.86	43.21
Corn Distillers Grains (pounds)	641.11	486.97	598.67
Dicalcium Phosphate (pounds)	—	40.97	—
Alfalfa Haylage (pounds)	772.24	1200.00	992.44
Cost per ton:	$ 45.91	$ 38.18	$ 41.31
Chemical Composition:			
Total pounds	2130.17	2067.59	2021.97
Crude Protein (pounds)	280.00	280.00	280.00
Crude Fiber (pounds)	360.00	360.00	360.00
Energy (Mcal/kg)	2.8	2.8	2.8
Calcium (pounds)	20	20	20
Phosphorus (pounds)	12	12	12

requirements. Throughout this book, the author has been careful to pay full tribute to the feeding standards and feeding programs now in use for the role they have played in bringing the dairy industry to its present high state of development. There is no reason to believe that current programs and feeding standards are incapable of supporting even further improvements.

Why, then, a new feeding program, and all of the current research and effort that has gone into its development. The reason is simple: the labor and individual feeding needed to utilize other feeding programs is not available for herds of the size we now have and will certainly not be available for the even larger herds of the future. The natural result is that those herds attempting to do some type of individual feeding under mass handling conditions are really doing a very poor job of feeding and are limiting the total production of their herds in the process. Thus, the concept of optimum rations which permit the individual cow to take care of her needs is almost a necessity if larger and larger herds are to achieve the level of milk production necessary to stay competitive in the dairy business. It is the author's considered opinion that on most large dairy farms there is not enough labor and individual attention to cows to permit herd averages above 12,000 to 13,000 pounds with conventional feeding systems.

Finally, the logical question. How does it work on dairy farms? What happens when a dairyman tests his silage, has a grain ration computed and proceeds to mix one combination of grain and silage and feed it as the only feed for his entire milking herd. We were fortunate in Georgia in having several dairy farmers who were willing to lead the way and try the program. Each of them submitted forage samples for analysis, we calculated a ration by one of the methods described in this chapter, and they fed the resulting mixture to their herds. Two 305-day DHIA herd averages for four such herds were:

Herd No.	Type of Cows	No. Cows	Milk	Fat
1	Holstein	50	16,308	600
	Holstein	45	17,098	610
2	Jersey	21	10,512	536
	Jersey	22	11,143	569
3	Holstein	132	14,216	540
	Holstein	146	13,215	524
4	Holstein	61	15,385	532
	Holstein	67	14,813	530

As with all feeding programs, the successful use of optimum rations depends upon the attention to details and a determination to make proper use of the information provided.

References

1. Hutton, R. F. Use of Linear Programming in Feed Manufacturing. A series of seven articles. Feedstuffs. September 27, October 4, October 11, October 18, October 25, November 1 and November 8, 1958 issues.

2. Taylor, R. D., Kohler, G. O., Maddy, K. H. and Enochian, R. V. Alfalfa Meal in Poultry Feeds—An Economic Evaluation Using Parametric Linear Programming. USDA, AEC Report No. 130. 1968.

Chapter 9

Principles of Meat Production

The ever-expanding number of people enjoying higher standards of living has put new pressures on the supply of beef throughout much of the world. This demand has played a major role in converting the bull calf born on dairy farms from an unwanted commodity to be disposed of as rapidly as possible to a valuable and profitable asset. Cull dairy cows and excess heifers that once were consigned to processed meats are now in great demand for crossbreeding to provide either brood cows for producing feeder calves or crossbred heifers for further crossing to produce fast gaining feeders. This changing picture has accompanied the rapid increase of Holstein cattle throughout North America and Western Europe. Dual purpose cattle and large breeds such as Brown Swiss have enjoyed increased demand due to the ability of their calves to produce rapid gains and heavy carcasses without excess fat.

In Western Europe, the use of bull calves as a meat source is an old practice. The thing that is new is the heavy demand and very high prices received by dairy farmers. In North America, the change is somewhat like turning back the clock to earlier times when beef was relatively unimportant and cows were kept as both a farm supply of milk and meat.

The rise in demand for dairy bulls for fattening and dairy blood for crossbreeding has necessitated a renewed interest in meat production on the part of dairy people. Thus dairy farmers are now concerned with all phases of ruminant nutrition including growth and fattening. This new interest in growth differs from older concerns since the emphasis is on maximum growth rates without concern for the fat deposition which may accompany it. The same people who are concerned with growing replacement heifers for milk production are now faced with learning new techniques of feeding since they are not concerned with the influence of early growth on later milk production.

WHAT IS GROWTH?

Growth has been defined as a correlated increase in mass of the body in definite intervals of time in a way characteristic of the species. Within such a complicated definition lies a major problem. The problem is that growth is sufficiently complicated and variable that it cannot be easily reduced to a simple set of causes and effects. When this situation exists in a production function, the door is wide open for everyone to have his own special system for doing the job. It is not surprising, therefore, that farm magazines run endless stories on the latest variation guaranteed to revolutionize the production of beef.

Despite the complexity of the problem, there are certain principles of growth which can be reduced to workable entities. Understanding these principles will go a long way toward removing the "side-show" atmosphere so common in the development of programs for beef production.

A first principle is that growth and fattening are not separate entities. The development of fatty tissue and the deposition of fat is a normal part of body development. Fat is not merely deposited as a result of feeding a ration in excess of maintenance requirements. Fat, once deposited, is not inert tissue that is stored awaiting a period of insufficiency when it can be used as a source of energy. The development of fatty tissue and the storage of fat must be considered as much a normal part of growth as the development of bones and muscles. The difference lies in priority. In normal growth, fatty tissue development has a low biological priority and thus develops more slowly than tissues with higher priorities. This principle is important, however, since it means that fattening can be accomplished at any stage of growth providing rations are consumed which permit even the low priority tissues to develop.

FACTORS AFFECTING GROWTH

The three primary factors affecting the growth characteristics of an animal are breed or type of animal, time, and nutrition. We will consider each in detail as a basis for understanding the production of beef.

Breed or Type of Animal One cannot divorce the importance of breed or type of animal from its performance and the desired end product. Thus, a late maturing but rapid gaining breed should be ideal for producing maximum meat providing the object is to provide maximum food without regard for any factor other than maximum conversion of nitrogen and calories into meat. If a meat quality factor is added (quality being defined as some ratio between protein and fat) then the mature weight and rate of gain must be juggled to meet the desired end product. Within a breed or type, the same decision must be made relative to the desired end product. In a wartime study in England, Watson (1943) found that cattle grown to produce calories should be slaughtered at about 1750 pounds live weight with a fat percentage of 35% and a killing-out percentage of 65%. If the same steers were valued only on protein, the optimum killing weight would have been 840 pounds with 22% fat and a killing-out percentage of 60%. However, if the beef was valued on the basis of consumer preference for a tasty combination, then an 1150 pound live weight, 24% fat, and a killing-out percentage of 59% was optimum.

Although the relative usefulness of any breed is determined by feed supplies and demands of the market, there are distinct characteristics which have been and will continue to be developed through breeding and selection.

Several variants due to breeding and selection were demonstrated in a California study shown in tables 56 and 57. The workers assembled seven lots of cattle representing distinct types of cattle. The longhorns were typical of cattle which predominated in the area in the late 1800s, the Angus and Herefords were selected to be representative of the breeds as were the Okies and Brahman. The Holsteins were steers from a large dairy herd and the 3-way crosses were from English x Dairy cows crossed with late maturing breeds. The steers were group-fed by breeds and slaughtered at comparable final weights. The data in table 56 indicate a wide range in average daily gains with the Brahman giving the lowest rate of gain followed by the longhorns while

Table 56 Influence of type of animal on average daily gain

Type	Initial weight	120-day ADG
Brahman	494	1.75
Longhorn	584	1.86
#1 Okies	644	2.24
Angus	568	2.42
Hereford	587	2.79
Holstein	581	3.22
3 x Cross	596	3.66

California data.

Table 57 Influence of type of animal on carcass characteristics

Type	Maturity score	Marbling score	Grade
Longhorn	14.2(A)	13.5(Sm—)	16.6(G)
Angus	14.5(A)	16.0(Mt)	19.7(C)
Hereford	14.2(A)	13.1(Sm)	18.7(C—)
#1 Okies	14.0(A)	13.2(Sm)	18.5(C—)
Brahman	14.5(A)	6.7(T—)	15.3(St)
Holstein	14.8(A)	11.2(S1)	16.4(G—)
3 x Cross	14.6(A—)	12.4(S1+)	18.1(G+)

California data.

the Holstein and 3-way crosses were tops for gain. The data in table 57 show the carcass characteristics of the animals. There was little difference in the maturity score, but vast differences in marbling and grade. The beef breeds showed the results of breeding and selection with their higher grades and the Angus the better marbling of earlier maturity. The Holstein and crosses indicated that the dairy breeds have also been bred for fast gains and earlier maturity—traits that have importance in getting heifers into production for milk. While there were considerable differences in what the carcasses looked like, there were much smaller differences in the total production of edible beef. This last factor is frequently overlooked in beef production. The glory frequently goes to those animals producing top grade carcasses and often overlooked is the contribution of other animals to a total meat supply. When the aim is to supply good nutrition to a population, the animal that will make rapid gains and efficient use of feed will be both the most productive and profitable. To accept anything less than this is to place a premium on consumer preference rather than nutrition. In an affluent society this may be a justifiable use of resources but in a society which needs good nutrition, it is a distinct waste of resources.

Time or Stage of Maturity As outlined in Chapter 15, growth and development is an orderly progression of events designed initially to prepare the animal for survival and ultimately for reproduction of the species. The general waves of growth have been outlined. It is important here to understand that the growth and development progresses in relation to the final weight and age of the animal at maturity.

In practice, the development of parts of the animal are important only in terms of the parts of the carcass that will develop most rapidly under a given level of nutrition at a stage of development of the animal. There are few detailed experiments in this field since the time and money required for this type of research has generally prohibited the use of large numbers of animals. Fortunately the general results have been in excellent agreement—thus reducing the need for duplicating research.

Overall, growth is a multiplication of tissues present at birth. The final size of the animal is the sum of these multiplications. On the average, the multiples of birth size of several tissues are: eye—2; brain—4; heart—12; liver—14; lungs—20; skeleton—26; muscles—48, and fat dependent upon degree of fattening.

One of the more extensive studies of these factors was conducted by Luitingh in South Africa. He used unfattened steers averaging 8, 20, and 32 months of age at the beginning of the study and fed representatives of each group at three levels of nutrition varying from a grain to roughage ratio of 1:3.5 to 2:1.

The data in tables 58 and 59 show some of the effects of stage of maturity at time of finishing on parts of the carcass. The young steers produced considerably less kidney fat and neck than the two year olds. The fact that the three year old cattle also produced a slower rate of development of the same two carcass parts indicated an advanced stage of development at the beginning

Table 58 Influence of age at fattening on growth rate of parts*

Part	8 mo. calf	2 year old	3 year old
Kidney fat	224	318	256
Neck	169	199	178
Shoulder	126	126	97
Prime rib	151	178	159
Loin	172	183	161

*g. day per 100 lb. of part.

Table 59 Influence of age at fattening on growth of organs and offal

Part	8 mo. calf	2 year old	3 year old
Caul fat	268	376	397
Liver	158	158	123
Stomach	158	113	67
Blood	126	122	87

of the test. Similar variations in shoulder, prime rib, and loin again indicated that each age group developed those parts which were ready or least developed at their stage of maturity. This trend is further demonstrated in the data in table 59. The older cattle deposited greater and greater quantities of cavel fat while the calves and two year olds completed the development of their vital organs and digestive tract.

In general, the overall effect of finishing the two and three year olds was to produce larger quantities of low priced cuts and offal. The data strongly support the feeding of younger cattle since the muscles and prime cuts apparently had at least equal call on the available nutrients with the vital organs and waste portions of the carcass.

The Influence of Level of Nutrition Most of the questionable "miracles" in growth and development have centered around the "special" ways for feeding cattle to produce carcasses with special attributes and to overcome the effects of type and age of cattle. Since these "miracles" seldom stand up under controlled research, the miracle-workers have an easy out through the same type and age factors they set out to overcome. The consensus of opinion is that within the limits of good rations normally fed, the level of nutrition plays a minor role in carcass characteristics outside of those determined by type and age of animal. In other words, so long as animals are maintained in positive energy balance and slaughtered at equal weights, the level of nutrition or combinations of feeding programs used to produce the carcass are relatively unimportant.

The data in tables 60 and 61 illustrate the results of an excellent experiment conducted by Lofgreen in California. He fed two ages of cattle—calves *vs.* yearlings and three energy levels—

Table 60 Nutritional treatments and carcass characteristics

(Calves)						
Treatment	Final wt.	Yield	Carcass wt.	Fat cover	Marbling score[1]	Final body fat
	(lb.)	(%)	(lb.)	(in.)		(%)
LLL	829	58.6	486	.51	5.0	18.2
LLH	911	61.8	563	.55	6.2	21.7
LMH	957	61.8	591	.63	6.6	22.5
HML	903	61.1	552	.71	6.0	20.5
HHL	907	61.5	558	.64	6.3	21.9
HHH	993	61.8	614	.73	6.3	25.3

[1]5 = small; 6 = modest; 7 = moderate.

Table 61 Nutritional treatments and carcass characteristics

(Yearlings)						
Treatment	Final wt.	Yield	Carcass wt.	Fat cover	Marbling score[1]	Final body fat
	(lb.)	(%)	(lb.)	(in.)		(%)
LLL	1036	59.9	621	.76	6.1	23.1
LLH	1113	61.4	683	.70	7.1	24.9
LMH	1223	64.1	784	.89	6.3	27.0
HML	1159	62.9	729	.90	7.3	26.2
HHL	1137	62.7	713	.88	7.3	27.0
HHH	1258	65.3	821	.97	7.2	29.4

[1]6 = modest; 7 = moderate; 8 = slightly abundant.

20% concentrate (L), 55% concentrate (M), and 90% concentrate to each group in three 90-day periods to provide six feeding combinations. The results have been interpreted in opposite ways by equally competent workers depending upon the treatment emphasized. The calves (table 60) fed the low level of nutrition throughout can be contrasted with those fed the high level to show a distinct effect of ration on carcass providing the final weight is ignored. Within the five levels which included a period on the high level of nutrition, there were small differences in all characteristics. Similar comments can be made for the data on the yearling steers (table 61). Overall, the age and weight of the cattle at slaughter resulted in the larger changes in marbling, fat cover, and percentage of fat in the carcass. The data does indicate that a leaner carcass with less fat could be produced by a very low level of nutrition of cattle throughout their lives. Such a program would result in the sale of old cattle produced at an enormous cost of nutrients and with carcasses typical of more mature steers. The value of such a practice brings us to the ultimate consideration of cost of production.

FACTORS AFFECTING COST OF BEEF PRODUCTION

Fundamental to a consideration of cost of meat production using cattle is an understanding of the factors which affects cost. Preston at the Ohio Research and Development Center divided the cost among the several nutrients as illustrated in table 62. Obviously, the relative percentages will change from time to time, but it is unlikely that the positions will change. The data indicates that the predominating cost is for energy. The relatively small percentage of the cost represented by protein, minerals, and vitamins would suggest that it would be difficult to find an economic condition when it would be advisable to feed these nutrients at less than an optimum level. If so, then the major decisions in deciding the most profitable nutritional level for feeding center around the cost of supplying energy.

In addition to the cost of feedstuffs, there are two animal factors which greatly influence the decisions on level and type of energy sources to be used in feeding all ruminants including those being used for meat production. The factors are interrelated and are the

Table 62 Relative cost of nutrients used in meat production

Nutrient	Requirement	% of Total Cost
Dig. Energy	26.7 Mcal	72.50
Crude Protein	950 gm	21.70
Calcium	26 gm	0.17
Phosphorus	20 gm	2.20
Vitamin A	25,000 IU	0.17
Salt	40 gm	0.42
Potassium	50 gm	1.70
Magnesium	5 gm	0.26
Sulfur	16 gm	0.75
Trace Min.	—	0.07

Based on an 800 lb. steer gaining 2.5 lb. at 1972 prices. (Preston)

low-level of intake of ruminants and the resulting high cost of maintenance.

In general, growing fattening steers will consume a maximum amount of dry matter as a proportion of body weight as follows:

Body weight	**Feed intake per 100 lbs.**
100	4.0
300	3.0
500	2.8
800	2.4
1000	2.2

Unlike monogastrics which will consume much larger quantities per unit of body weight, the ruminant operates at a low level of efficiency due to the large percentage of the total ration going into maintenance. Thus, intake of total dry matter must rate as a major factor in limiting the efficiency of feed conversion for ruminants.

This factor can be partially overcome by increasing the energy concentration per unit of feed, but as explained in Chapter 4, this route has limitations due to the drop in intake on rations containing over 60% concentrates. The data in figure 38 illustrate this point. The dotted lines illustrate the theoretical reduction in cost per pound of gain that could result from increasing concentration

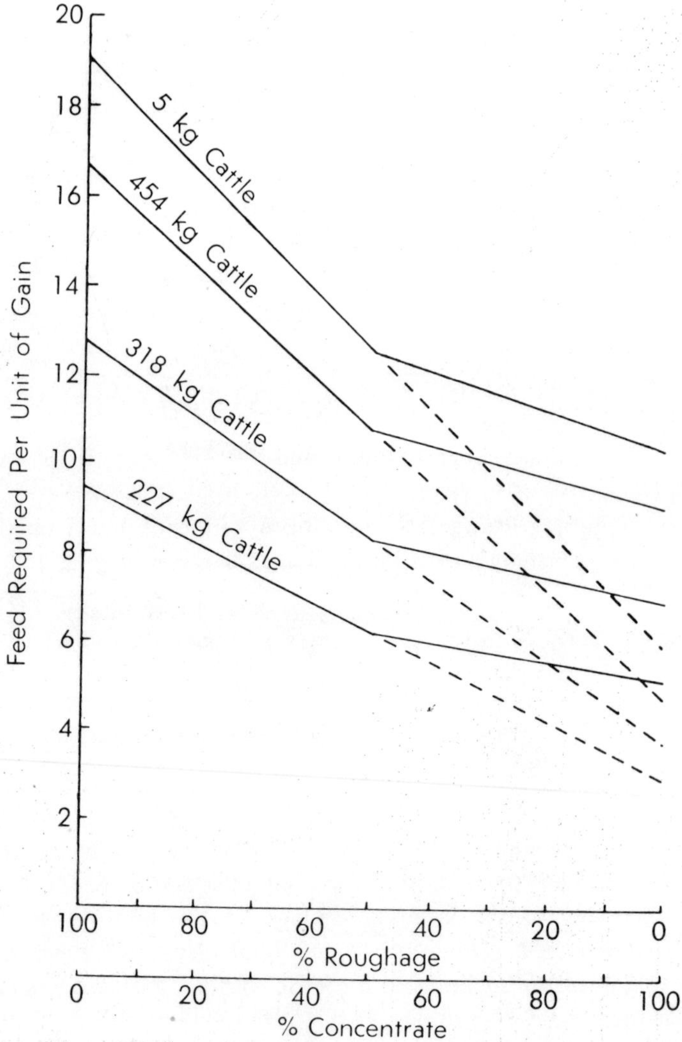

Figure 38 Expected feed efficiency of cattle fed varying proportions of roughage and concentrate. (Preston, Ohio State)

of the ration while the solid lines illustrate the expected change due to this intake limitation. Thus, any great hope for making a significant increase in feed utilization must stem from the ability to increase energy intake in the form of high energy rations. Since this must be done within the limitations of rumen fermentation the problem will be a complex one.

The influence of this intake limitation in combination with the high cost of body maintenance is illustrated in table 63. To produce 600 pounds of gain at 1.5 pounds ADG requires a feeding period of 400 days and utilizes more total feed for maintenance than for gain. Feeding for 2.8 pounds of gain per day reduces total feed required by about 40% and that used for maintenance by about 50%. These values would vary depending upon the size and age of the animal being fed. The change, however, is sufficiently large enough that it will overcome vast differences in cost of ingredients required to produce the higher rate of gain. This means in practice that over a wide range of cost conditions, animals are probably best fed for a near maximum rate of gain

Table 63 Use of feed and protein for meat production as influenced by rate of gains

	Pounds dry matter for			
	Maintenance		Gain	
ADG	Feed	Protein	Feed	Protein
1.50	3387	389	3128	355
2.20	2004	233	2450	285
2.80	1545	189	2417	296

There is, of course, a difference between feeders who buy all their feed and the farmer feeder who used animals either to market otherwise non-marketable forage or to increase the market value of the crops produced on his farm. In the latter case, the total beef produced per acre may be more important than rate of gain or per animal performance. So long as it is understood that this difference in measures of success does not alter the basic nutritional relationships, valid economic decisions can be made which include this difference in objectives. Whether one buys his feed or grows it on his farm does not alter the fact that, at low levels of performance, a major portion of the feed consumed goes into body maintenance which cannot be sold and a small proportion into gain which can be sold. In either case, the enterprise will be only as profitable as the income above feed cost permits.

References

1. Adams, Nelson J. and Frank Fox. Performance and Carcass Characteristics of Various Types of Cattle Fed for Beef. California Livestock Symposium. June, 1972.
2. Liutingh, H. C. Developmental Changes in Beef Steers as Influenced by Fattening, Age, and Type of Ration. J. Agric. Sci. 58:1. 1962.
3. Lofgreen, G. P. The Effect of Nutrition on Carcass Characteristics. 4th Annual Arizona Feeds—Elanco Seminar. 1968.
4. Preston, R. L. Nutritional Implications in Economy of Gain of Feedlot Cattle. J. Animal Sci. 35:153. 1972.
5. Watson, D. M. S. Beef Cattle in Peace and War. Empire J. Exp. Agric. 11:191. 1943.

Appendix

Ingredient Composition, Ration Requirements and Restrictions for Linear Programmed Rations

Least cost linear programmed rations are generally considered only as a means of determining the least cost rations from available feed supplies. In many feeding situations, the available feedstuffs are determined by previous farming decisions and the need is frequently for a convenient method for calculating a balanced ration. Ration balancing or feed programming has become increasingly difficult due to the desirability of considering several minerals as well as the traditional energy and protein. The program presented here is equally useful for either purpose.

From a nutritional point of view, the program includes the following assumptions and characteristics:

1. The program includes requirements for crude protein, crude fiber (milk production only), energy (either as net or metabolizable), roughage value, calcium, phosphorus, magnesium, and potassium. The values are in line with NRC and ARC recommendations, but have been changed where necessary to conform with current research and field experience.

2. The ration requirements are based on normal levels of intake of good to excellent forages such as hay or silage. When feedstuffs which contribute only roughage are used, it may be necessary to remove the roughage restrictions from the rations.

3. Assuming normal levels of dry matter intake, the rations provide optimum conditions for the level of performance indicated. If intake varies from the assumed levels, the rations remain in nutrient balance and performance will reflect energy intake. Since animal performance is dependent upon factors such as type of animal, previous nutritional history, age of animal, as well as feedstuffs used, the concept of feeding optimum rations serves well to insure the intake of a balanced ration regardless of level of performance. In practice, it is easier and more realistic to make adjustments in the recommended ration to fit it to the specific herd or lot of cattle than to attempt to provide specifications for specific rations and guess which group of specifications fits the situation. Regardless of the system used to select the ration, fitting the ration to the situation requires a certain amount of trial and error with the animals being fed.

In terms of technique, the program provides the following possibilities:

1. At the Georgia Station, the program is used on an IBM 1620, Mod 1, 60K machine with a 1443 printer and two 1311 disc packs. The solution algorithm is referred to as the IBM 1620-1311 Linear Programming system. The source language used for the 1620-1311 Linear Programming system is IBM 1620 SPS II-D. The Linear Programming program is run under control of IBM 1620 Monitor I system.

2. Cost ranges are calculated only for ingredients that are in the formulation at an intermediate level. Cost ranges are not provided for ingredients forced into the solution at a specific level or for ingredients included in the formulation at an upper or lower bound. Cost ranges are not available for ingredients not in the rations. The information provided by the cost range is useful in determining the relative stability of a least cost formulation. If the range in prices for all ingredients is narrow, the chances of this particular formulation remaining at least cost level for very long will be small. Actually, information provided by cost ranges is limited. Changes can only be interpreted for price variations of one ingredient at a time.

Information concerning ingredients not contained in the least cost formulation and ingredients at upper and lower bounds is referred to as reduced cost. For excluded ingredients it is possible to determine the price at which they would enter the ration.

If an ingredient is contained in the formulation at its upper bound (the maximum level as permitted in the ration), the economic interpretation of the reduced cost is the reduction in the cost of the ration if the restriction were relaxed and one more unit of the variable were allowed to enter. For ingredients at a lower limit (the minimum level as permitted in the ration) or that are forced into the ration at a specific level, the interpretation of the reduced cost is the cost of forcing one more unit of variable into the solution by increasing the limit.

The reduced cost also allows the calculation of the highest feasible price that could be paid for an excluded ingredient so that it would come into the ration without increasing the cost of the ration.

The following tables present the data used for ingredient composition and restrictions, expected animal performance and ration characteristics and restrictions for milk production and growth and fattening of cattle.

Table 64 Expected Animal Performance

Rations Class	**Number**
Milk Production	1 - Optimum (NE_{milk})
	2 - < 40 lb. "
	3 - 40-60 lb. "
	4 - > 60 lb. "
	5 - Optimum (Met. Energy)
	6 - < 40 lb. "
	7 - 40-60 lb. "
	8 - > 60 lb. "
Growth and Fattening	9 - Young heifers (replacements)
	10 - Older heifers (replacements)
	11 - 200-300 lbs. (2.00 lb. ADG)
	12 - 200-300 lbs. (2.50 lb. ADG)
	13 - 200-300 lbs. (2.75 lb. ADG)
	14 - 300-500 lbs. (2.00 lb. ADG)
	15 - 300-500 lbs. (2.50 lb. ADG)
	16 - 300-500 lbs. (2.75 lb. ADG)
	17 - 500-700 lbs. (2.00 lb. ADG)
	18 - 500-700 lbs. (2.50 lb. ADG)
	19 - 500-700 lbs. (2.75 lb. ADG)
	20 - 700-900 lbs. (2.00 lb. ADG)
	21 - 700-900 lbs. (2.50 lb. ADG)
	22 - 700-900 lbs. (2.75 lb. ADG)
	23 - 900-1100 lbs. (2.00 lb. ADG)
	24 - 900-1100 lbs. (2.50 lb. ADG)
	25 - 900-1100 lbs. (2.75 lb. ADG)

Table 65 Composition, roughage factor and restrictions for feed ingredients

Ingredient	Crude protein	Crude fiber	Calcium	Phosphorus	Magnesium	Potassium	Mcal NE_{milk}	Mcal NE_{gain}	Metabolizable energy (Mcal)	Roughage factor	Restriction % of ration Milk	G&F
	per pound										(Maximum)	
Alfalfa Meal, DEHY	.170	.250	.0130	.0023	.0029	.0250	0.59	0.29	0.94	0.20	5.0	5.0
Animal Fat	.000	.000	.0000	.0000	.0000	.0000	0.99	1.18	1.42	0.00	2.0	2.0
Barley	.115	.070	.0006	.0036	.0012	.0056	0.86	0.56	1.25	0.10	25.0	0.0
Beet Pulp, dried	.090	.190	.0060	.0010	.0027	.0016	0.73	0.42	1.09	0.05	15.0	25.0
Brewers Grains, dried	.259	.150	.0027	.0050	.0019	.0006	0.65	0.35	1.00	0.20	0.0	10.0
Citrus Pulp, dried	.066	.130	.0197	.0012	.0016	.0062	0.80	0.54	1.17	0.20	0.0	25.0
Corn	.090	.020	.0002	.0029	.0015	.0033	0.98	0.60	1.38	0.00	0.0	0.0
Corn and Cob Meal	.081	.080	.0004	.0027	.0015	.0053	0.87	0.54	1.26	0.20	0.0	0.0
Corn Cobs	.023	.350	.0012	.0004	.0016	.0076	0.37	0.10	0.66	0.60	0.0	0.0
Corn Gluten Feed	.220	.090	.0046	.0077	.0029	.0060	0.86	0.53	1.25	0.00	0.0	0.0
Cottonseed Meal (Expeller)	.410	.160	.0015	.0100	.0054	.0120	0.98	0.45	1.47	0.00	0.0	0.0
Cottonseed Meal (Solvent)	.420	.120	.0015	.0110	.0058	.0147	0.77	0.46	1.15	0.00	0.0	0.0
Cottonseed Hulls	.040	.430	.0014	.0009	.0013	.0078	0.31	0.04	0.60	1.00	20.0	0.0
Distillers Dried Grains	.270	.120	.0009	.0037	.0006	.0009	0.91	0.55	1.31	0.10	0.0	0.0
Distillers Dried Grains/solubles	.270	.085	.0012	.0068	.0006	.0009	0.97	0.59	1.38	0.10	0.0	0.0
Distillers Dried Solubles	.270	.040	.0035	.0130	.0064	.0174	0.97	0.59	1.38	0.00	0.0	0.0
Hominy Feed, Yellow	.105	.050	.0005	.0053	.0024	.0067	1.04	0.63	1.47	0.00	0.0	0.0
Linseed Meal (Expeller)	.330	.090	.0044	.0089	.0058	.0124	0.85	0.52	1.25	0.00	0.0	0.0
Linseed Meal (Solvent)	.340	.090	.0040	.0083	.0060	.0138	0.79	0.47	1.16	0.00	0.0	0.0
Malt Sprouts, Dried	.250	.150	.0020	.0070	.0018	.0021	0.65	0.35	1.02	0.20	10.0	10.0

Table 65 Continued

Ingredient	Crude protein	Crude fiber	Calcium	Phos-phorus	Mag-nesium	Potas-sium	Mcal NE_{milk}	Mcal NE_{gain}	Metabolizable energy (Mcal)	Roughage factor	Restriction % of ration Milk (Maximum)	G&F
	— — —	— — —	— — —	— — —	— — —	per pound	— — —	— — —	— — —	— — —		
Milo, Grain Sorghum	.090	.025	.0004	.0031	.0017	.0034	0.86	0.53	1.25	0.00	0.0	0.0
Molasses, Beet	.065	.000	.0010	.0002	.0023	.0477	0.66	0.40	0.98	0.00	10.0	10.0
Molasses, Cane	.030	.000	.0082	.0008	.0035	.0238	0.82	0.50	1.17	0.00	10.0	10.0
Oats	.120	.120	.0010	.0013	.0017	.0037	0.77	0.46	1.14	0.10	15.0	15.0
Oat Hulls	.040	.300	.0009	.0010	.0019	.0040	0.31	0.04	0.60	1.00	0.0	0.0
Peanut Hulls	.080	.630	.0025	.0006	.0019	.0082	0.16	0.00	0.29	1.00	0.0	0.0
Peanut Meal (Expeller)	.450	.120	.0015	.0055	.0033	.0115	0.89	0.55	1.29	0.00	0.0	0.0
Peanut Meal (Solvent)	.470	.130	.0020	.0060	.0004	.0115	0.73	0.48	1.19	0.00	0.0	0.0
Rice Bran	.130	.120	.0010	.0150	.0094	.0135	0.63	0.35	0.99	0.15	5.0	5.0
Rice Hulls	.044	.400	.0008	.0006	.0010	.0034	0.16	0.00	0.29	1.00	0.0	0.0
Soybeans	.380	.050	.0025	.0059	.0021	.0170	1.03	0.62	1.45	0.00	5.0	5.0
Soybean Hulls	.125	.350	.0054	.0016	.0070	.0086	0.90	0.49	1.20	0.00	20.0	20.0
Soybean Meal (Expeller)	.420	.065	.0020	.0060	.0027	.0197	0.90	0.56	1.30	0.00	0.0	0.0
Soybean Meal (Solvent)	.440	.070	.0025	.0060	.0027	.0197	0.83	0.52	1.22	0.00	0.0	0.0
Soybean Meal (Solvent) dehulled	.490	.025	.0020	.0065	.0027	.0197	0.90	0.56	1.30	0.00	0.0	0.0
Urea	2.820	.000	.0000	.0000	.0000	.0000	0.00	0.00	0.00	0.00	0.5	1.0
Wheat, Hard	.130	.020	.0005	.0040	.0011	.0032	0.90	0.56	1.29	0.00	0.0	0.0
Wheat, Soft	.100	.028	.0005	.0030	.0010	.0041	0.90	0.56	1.29	0.00	0.0	0.0
Wheat Bran	.160	.100	.0014	.0117	.0045	.0120	0.68	0.38	1.03	0.00	0.0	0.0
Wheat Middlings	.170	.080	.0015	.0091	.0008	.0034	0.0087	0.53	1.26	0.00	0.0	0.0
Wheat Shorts	.184	.050	.0011	.0076	.0026	.0088	0.0091	0.56	1.31	0.00	0.0	0.0
Silage					use forage test					1.00	60.0	0.0

Table 65 Continued

Ingredient	Crude protein	Crude fiber	Calcium	Phos-phorus	Mag-nesium	Potas-sium	Mcal NE_{milk}	Mcal NE_{gain}	Metabolizable energy (Mcal)	Roughage factor	Restriction % of ration	
	per pound										Milk	G&F
											(Maximum)	
Bone Meal, Steamed	.000	.000	.2900	.1400	.0000	.0000	.0000	0.00	0.00	0.00	0.0	0.0
Defluorinated Phosphate	.000	.000	.3400	.1800	.0000	.0000	.0000	0.00	0.00	0.00	0.0	0.0
Dicalcium Phosphate	.000	.000	.2300	.1800	.0000	.0000	.0000	0.00	0.00	0.00	0.0	0.0
Limestone	.000	.000	.3800	.0000	.0000	.0000	.0000	0.00	0.00	0.00	0.0	0.0
Magnesium Carbonate	.000	.000	.0000	.0000	.2600	.0000	.0000	0.00	0.00	0.00	0.0	0.0
Monosodium Phosphate	.000	.000	.0000	.2600	.0000	.0000	.0000	0.00	0.00	0.00	0.0	0.0
Double Sulf. of Pot.-Mg.	.000	.000	.0000	.0000	.1100	.1800	.0000	0.00	0.00	0.00	0.0	0.0

Table 66 Rations for milk production (net energy)

Specification	Unit	Restrictions	Ration 1 Optimum*	Ration 2 < 40 lbs.	Ration 3 40-60 lbs.	Ration 4 > 60 lbs.
Quantity	pounds	equality	2000	2000	2000	2000
Cost	dollars	minimum	?	?	?	?
Crude protein	lb/ton	minimum	260	280	300	320
		maximum	280	290	310	330
Crude fiber	lb/ton	minimum	320	320	320	300
		maximum	400	400	400	350
NE_{milk}	Mcal/ton	minimum	1500	1270	1450	1632
Calcium	lb/ton	equality	14	14	14	14
Phosphorus	lb/ton	equality	10	10	10	10
Magnesium	lb/ton	minimum	4	4	4	4
Potassium	lb/ton	minimum	16	16	16	16
T/M salt	lb/ton	equality	10	10	10	10
Roughage factor	units/ton	minimum	800	1000	800	600

*Optimum—to be fed as a single ration to all cows.

Table 67 Rations for milk production (metabolizable energy)

Specification	Unit	Restrictions	Ration 5 Optimum	Ration 6 < 40 lbs.	Ration 7 40-60 lbs.	Ration 8 > 60 lbs.
Quantity	pounds	equality	2000	2000	2000	2000
Cost	dollars	minimum	?	?	?	?
Crude protein	lb/ton	minimum	260	280	300	320
		maximum	280	290	310	330
Crude fiber	lb/ton	minimum	320	320	320	300
		maximum	400	400	400	350
Met. Energy	Mcal/ton	minimum	2300	1995	2358	2600
Calcium	lb/ton	equality	14	14	14	14
Phosphorus	lb/ton	equality	10	10	10	10
Magnesium	lb/ton	minimum	-4	4	4	4
Potassium	lb/ton	minimum	16	16	16	16
T/M Salt	lb/ton	equality	10	10	10	10
Roughage factor	units/ton	minimum	800	1000	800	600

Table 68 Rations for growth and fattening (net energy)

Ration Type: Specifications	Unit	Restrictions	Young heifers 9	Older heifers 10	200-300 lb 11	200-300 lb 12	200-300 lb 13
Quantity	pounds	equality	2000	2000	2000	2000	2000
Cost	dollars	minimum	?	?	?	?	?
Crude protein	lbs/ton	equality	280	220	260	280	300
NE_{gain}	Mcal/ton	minimum	900	800	1000	1060	1120
Calcium	lbs/ton	minimum	10	10	10	10	10
		maximum	12	12	12	12	12
Phosphorus	lbs/ton	minimum	8	8	8	8	8
		maximum	10	10	10	10	10
Magnesium	lb/ton	minimum	4	4	4	4	4
Potassium	lb/ton	minimum	14	14	14	14	14
T/M Salt	lb/ton	equality	10	10	10	10	10
Roughage factor	units/ton	minimum	800	800	400	400	200

Table 69 Rations for growth and fattening (metabolizable energy)

Ration Type: Specification	Unit	Restrictions	200-300 lb 26	200-300 lb 27	200-300 lb 28	300-500 lb 29	300-500 lb 30	300-500 lb 31
Quantity	pounds	equality	2000	2000	2000	2000	2000	2000
Cost	dollars	minimum	?	?	?	?	?	?
Crude protein	lbs/ton	equality	260	280	300	240	260	280
Met. Energy	Mcal/ton	minimum	2440	2500	2560	2440	2500	2560
Calcium	lbs/ton	minimum	10	10	10	10	10	10
		maximum	12	12	12	12	12	12
Phosphorus	lbs/ton	minimum	8	8	8	8	8	8
		maximum	10	10	10	10	10	10
Magnesium	lbs/ton	minimum	4	4	4	4	4	4
Potassium	lbs/ton	minimum	14	14	14	14	14	14
T/M Salt	lbs/ton	equality	10	10	10	10	10	10
Roughage factor	units/ton	minimum	400	400	200	400	400	400

300-500 lb			500-700 lb			700-900 lb			900-1100 lb		
14	15	16	17	18	19	20	21	22	23	24	25
2000	2000	2000	2000	2000	2000	2000	2000	2000	2000	2000	2000
?	?	?	?	?	?	?	?	?	?	?	?
240	260	280	240	240	240	240	240	240	220	220	220
1000	1060	1120	960	1030	1100	940	1020	1100	960	1030	1100
10	10	10	9	9	9	9	9	9	6	6	6
12	12	12	10	10	10	10	10	10	9	9	9
8	8	8	7	7	7	7	7	7	5	5	5
10	10	10	8	8	8	8	8	8	8	8	8
4	4	4	4	4	4	4	4	4	4	4	4
14	14	14	14	14	14	14	14	14	14	14	14
10	10	10	10	10	10	10	10	10	10	10	10
400	400	200	400	400	400	400	400	400	400	400	400

500-700 lb			700-900 lb			900-1100 lb		
32	33	34	35	36	37	38	39	40
2000	2000	2000	2000	2000	2000	2000	2000	2000
?	?	?	?	?	?	?	?	?
240	240	240	240	240	240	220	220	220
2400	2480	2560	2360	2460	2560	2400	2500	2600
9	9	9	9	9	9	6	6	6
10	10	10	10	10	10	9	9	9
7	7	7	7	7	7	5	5	5
8	8	8	8	8	8	8	8	8
4	4	4	4	4	4	4	4	4
14	14	14	14	14	14	14	14	14
10	10	10	10	10	10	10	10	10
400	400	400	400	400	400	400	400	400

Index